APPLIED MECHANICS

PRENTICE-HALL INTERNATIONAL, INC., *London*
PRENTICE-HALL OF AUSTRALIA, PTY. LTD., *Sydney*
PRENTICE-HALL OF CANADA, LTD., *Toronto*
PRENTICE-HALL OF INDIA PRIVATE LIMITED, *New Delhi*
PRENTICE-HALL OF JAPAN, INC., *Tokyo*

APPLIED
MECHANICS

G. WAYNE BROWN

Cambrian College of
Applied Arts and Technology

PRENTICE-HALL, INC., Englewood Cliffs, N.J.

Current printing (last digit):
10 9 8 7 6 5 4 3

13-041301-1

Library of Congress Catalog Card Number: 70-145957
Printed in the United States of America

PREFACE

The treatment of Applied Mechanics in this book is quite traditional, yet different from that used by most others who have written for the technician, the technologist, or the engineer. The differences between this and other similar works might be most apparent if the objectives I used in writing this book are enumerated as follows:

1. To write for the student. Concepts and ideas are frequently expressed in several different ways in the hope that the student will have a better chance of understanding the material. Those who use this book as instructors may find the rephrasing of concepts unnecessarily repetitive, but the book is primarily for those learning the concepts for the first time.

2. To present both graphical and analytical methods of problem solution where possible. The book has been written so that it is possible to place emphasis on either approach, although it is my conviction that one should develop competence in both methods of problem solution.

3. To emphasize the use of vectors. This is not carried to the extreme of using vector algebra, but the reader is regularly reminded that he is using quantities which must be described by both a magnitude and a direction.

4. To treat Applied Mechanics as a reasoning process, and not just a collection of formulae to be memorized. In

v

most topics, the general case is discussed. Thus the student is not burdened with trying to match a large number of formulae with a large number of special conditions.

5. To provide adequate explanation in example problem solutions. The format used for example problems is different from that used elsewhere. An attempt is made to let the reader see all the reasoning used in arriving at a particular solution.

6. To provide realistic problems. As far as possible, the emphasis in problem selection has been on the "applied" of Applied Mechanics.

Discussion and comment from users of this book would be most welcome. Having enumerated my objectives, it would be very valuable to find out how well others consider the objectives to have been met. Comments from student users are particularly welcome. They may be directed to me at the publisher's address.

In any task such as the preparation of a book, many people provide assistance. The most valuable assistance has come from the many students who have been using this book in draft form, and contributing to it by giving their comments, criticism, and most important, their encouragement. Special thanks goes to Larry McIntosh, Karl Vainio, and Don Cook who have read the final manuscript to provide a final student review. The constructive criticisms of the publisher's reviewer, and of Professor A. S. Weaver, of Michigan Technological University, are also appreciated. My wife has given valuable assistance in proofreading, and my children must also be thanked for being so understanding about the lack of attention they have received while the book was being prepared.

G. WAYNE BROWN

CONTENTS

Chapter **4** EQUILIBRIUM 55

Chapter **5** THREE-DIMENSIONAL STATICS 107

Chapter **6** FRICTION 133

APPLIED MECHANICS

1

INTRODUCTION

Applied mechanics is referred to by a number of different names. It may be called simply *mechanics*, or *rigid body mechanics*, or *Newtonian mechanics*, or perhaps *engineering mechanics*. It might even be called *engineering science*. Regardless of the name, the material covered will be about the same, and the emphasis will be on the application of a few principles of that branch of physics called mechanics.

Before reading further, we would like to ask you to go back and read the preface of this book if you have not already done so. One of the functions of a preface is to tell the readers about the book in a general or nontechnical fashion. Reading the preface of this, or almost any text, will give you an understanding of the kind of knowledge you are to acquire and the way it is to be acquired.

1-1. WHAT IS APPLIED MECHANICS

As stated above, applied mechanics is simply the study of the application of a few principles from physics. Since you probably know the principles, Newton's laws, already, we arrive at our first problem. Although many students recognize that the principles of equilibrium and motion have been dis-

1

cussed in prior studies of physics, they assume that they are not studying anything new or different in applied mechanics. However, the emphasis is on the application of these principles. Much diligent practice will be required to develop the facility needed to solve the great variety of problems encountered in this book and in the many areas where applied mechanics is used.

For those who like definitions, we might define applied mechanics as a study of the application of Newton's laws to find forces on bodies and to describe the motion of bodies. Frequently, the subject is divided into two parts—*statics* and *dynamics*. Statics is the study of forces acting on bodies at rest or at constant speed in a straight line, and dynamics is the study of the geometry of motion and the forces causing motion.

1-2. VALUE OF APPLIED MECHANICS

Every student should ask why he is studying a subject. Since there is so much knowledge available and so little time to learn it, one can be reasonably certain that there is time to study only important subject matter. Applied mechanics, like many other subjects, is not an end in itself. Its value is in the background it gives the student in preparation for other types of problems.

The first benefit from a study of applied mechanics is rather subtle. It develops your skill in analyzing or determining how to approach the solution of a problem. A more concrete value of applied mechanics is its relationship to other areas of study. For example, the design of any type of structure depends on your ability in applied mechanics; before designing the structure, you must first determine the loads on each part of the structure. This is a problem in applied mechanics. The problems of fluids, whether you think in terms of an hydraulic system or an airplane moving through the air (a fluid), require an ability in fluid mechanics, which in turn requires a working knowledge of statics and dynamics.

Many illustrations of the use of applied mechanics appear in Fig. 1-1, which shows an aircraft being loaded with freight. The design of the airplane, both as a flying machine and as a structure, depends on the use of the principles of

Fig. 1-1. The aircraft, the fork lift truck, and even the pavement are designed using the principles of applied mechanics. (Photo courtesy of Air Canada.)

applied mechanics. The problems of propulsion, lift, and directional control are fluid mechanics problems, for the air in which the plane operates is a compressible fluid. One can study the behavior of bodies in such a fluid only after he has mastered the principles of statics and dynamics.

If we define a structure as any system which must support a load, then the airplane is a structure. One cannot design such a structure until he has mastered the principles of statics and dynamics sufficiently well in order to determine the forces applied to each part of the structure due to both the dead loads—the aircraft's weight and the cargo—and the live loads that result from acceleration, takeoff and landing, and wind.

There are other mechanics problems, again associated with fluid mechanics, in the design of the hydraulic control systems which operate the landing gear, flaps, and other control surfaces. Of course, engines have just as many problems for they contain many moving parts whose motions must be analyzed so that the parts do not interfere with one another. The engine parts, too, have forces exerted on them which must be calculated so that the parts can be made of appropriate size and material to prevent failure.

The problems in designing the lift truck are similar, although not quite so complex. At least in the lift truck, aerodynamic behavior is not very important.

We have not yet approached the end of the list of items whose design relies upon the use of the principles of applied mechanics. The pavement on which the aircraft and lift truck operate must be carefully designed so that it is strong enough to support these vehicles, particularly under the extremely high forces that can occur when the aircraft lands. Even the soil supporting the pavement must be investigated. Again the principles of applied mechanics are used to determine if the soil is strong enough to carry the loads to be imposed.

When you have completed this study of applied mechanics, you will not be ready to design such complicated mechanisms as aircraft. However, you will be well prepared to continue your studies in fluid mechanics, stress analysis, machine design, and structural design, so that ultimately you will be prepared to assist in the design of a great variety of devices for mankind, ranging all the way from pipelines for fresh water to rockets.

1-3. HISTORY

Scientific facts have generally existed for a long, long time. However, when we look at the history of science, we are concerned with the time when man observed these facts and recognized their order. The earliest known discussion of mechanics was by Archimedes, a Greek, who lived about 250 B.C. He stated the principle of buoyancy and explained the use of levers. There was little advance in knowledge of the principles of mechanics for many years thereafter. In fact, knowledge of most sciences died out in the western world with the fall of the Roman Empire.

It was not until the beginning of the seventeenth century that the next significant development occurred. At that time, Simon Stevin, a Dutchman, developed the principles of vector addition, which made it possible to express many of the principles of statics in mathematical form. During the next century, Galileo performed some experiments in dynamics, with his study of falling bodies, and Isaac Newton expressed his three laws of motion on which our study of mechanics is based.

Although our study of mechanics will be based on Newton's laws, new theories of mechanics are still being developed. One such theory is Einstein's *relativistic mechanics*, which deals with the motion of bodies traveling at or near the speed of light. *Quantum mechanics* is another modern specialized field of mechanics. Most of the technical problems which we will encounter can be solved with Newtonian mechanics, so that we will not have need to study the other types of mechanics.

2

TOOLS FOR APPLIED MECHANICS

The emphasis throughout this book will be on developing your ability to solve problems using the few basic principles of mechanics which will be introduced. The many problems will help to develop your skill in technical analysis. You will find that remembering a formula is not adequate for solving problems. The challenge will be in determining how to attack the problem, and making use of the few formulas you need to arrive at the correct solution.

2-1. METHODS OF PROBLEM SOLUTION

There are two basic methods for solving most problems. These are the *analytical* and the *graphical* methods of solution. You can find people who will argue over which is the best method of solution, but we think there is no best method. Sometimes the graphical method provides the quickest means of obtaining a reasonably accurate solution, and sometimes the analytical method is quickest and most accurate. Where possible, the best method is to solve a problem using either the analytical or graphical method, whichever you prefer, and then use the other method to solve the problem a second time so that you can check your work. Developing an ability to check your work is important, for in industry, you will quickly discover that no one provides you with a list of answers to your problems such as the list in the back of this book.

Regardless of how you decide to solve your problems, there is another habit you should form as soon as possible. It is the habit of doing neat, orderly work, with adequate explanation for each step. Start now to do all work as if you will be expected to hold it up for the world to see. Everything you do should be done so well that you will be proud to show it to instructors, coworkers, or supervisors. The habit of excellent work will benefit you in the classroom or on the job. It will mark you as a person who cares about the kind of work he does and will improve your chances for good grades in the class and advancement on the job.

A student working on applied mechanics problems is shown in Fig. 2-1. Notice that he has adequate work space and all the equipment at hand to solve his problems both graphically and analytically.

2-2. ANALYTICAL METHOD

The procedure of describing force or motion relationships by mathematical expressions, and using these expressions to solve for the unknown quantities we are seeking, is known as the analytical method. The mathematics you will require includes ordinary arithmetic, some algebra, and some trigonometry.

The algebra which you need most is the ability to solve linear equations with one unknown. In some cases you will need to solve simultaneous linear equations—usually two equations with two unknowns. The other bits of knowledge you will use on some occasions is the quadratic formula, or factoring, for finding values of an unknown when the unknown is squared.

Trigonometry is the study of the relationship between angles and sides in a triangle. The laws and rules of trigonometry are all based on the definitions of *trigonometric functions* (usually called *trig functions*) listed below. These definitions refer to the right-angled triangles shown in Fig. 2-2(a). Angles are usually designated by Greek letters, whose names and shapes can be found in a mathematics handbook.

Fig. 2-1. A student is shown solving problems under near ideal conditions. He has plenty of room, and all equipment near at hand for solving problems both graphically and analytically. (Photo courtesy of Karl Sommerer and Cambrian College of Applied Arts and Technology.)

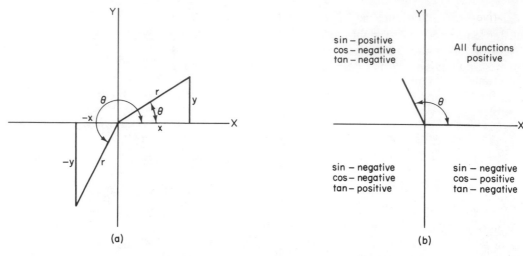

Figure 2-2

The angles are considered to be positive if they are formed by counterclockwise rotation from an axis.

The short forms for the names of the trig functions are generally used. So that you will be familiar with the full names, they are shown in the list below.

$$\text{sine } \theta = \sin \theta = \frac{\text{opposite}}{\text{hypotenuse}} = \frac{y}{r}$$

$$\text{cosine } \theta = \cos \theta = \frac{\text{adjacent}}{\text{hypotenuse}} = \frac{x}{r}$$

$$\text{tangent } \theta = \tan \theta = \frac{\text{opposite}}{\text{adjacent}} = \frac{y}{x}$$

$$\text{cosecant } \theta = \csc \theta = \frac{\text{hypotenuse}}{\text{opposite}} = \frac{r}{y}$$

$$\text{secant } \theta = \sec \theta = \frac{\text{hypotenuse}}{\text{adjacent}} = \frac{r}{x}$$

$$\text{cotangent } \theta = \cot \theta = \frac{\text{adjacent}}{\text{opposite}} = \frac{x}{y}$$

where

opposite represents the side opposite the angle, θ,
adjacent represents the side adjacent to the angle, θ, and
hypotenuse is the hypotenuse of the right-angled triangle.

It should be noted that x and y will have signs associated with them, depending on whether they are directed toward the positive or negative direction of the axes. The hypotenuse r is always positive. You can see that it follows that the trig functions will also have signs depending on the signs of x and y. Fig. 2-2(b) shows the signs for the most important trig functions in the four quadrants. The relationships between trig functions for angles in the first and second quadrants are as follows:

$$\sin \theta = \sin (180° - \theta) = -\cos (90° + \theta)$$
$$\cos \theta = -\cos (180° - \theta) = \sin (90° + \theta)$$
$$\tan \theta = -\tan (180° - \theta) = -\cot (90° + \theta)$$

Values of trig functions can be obtained from a book of math tables or from your slide rule. In either case you will have to supply the correct sign.

Of course, all triangles are not right angled. To deal with triangles without a right angle there are two laws, known as the *sine law* and the *cosine law*. These relate the angles and sides for triangles which do not have right angles. The sine law is

$$\frac{\sin \alpha}{A} = \frac{\sin \beta}{B} = \frac{\sin \gamma}{C} \tag{2-1}$$

The cosine law is

$$C^2 = A^2 + B^2 - 2AB \cos \gamma \tag{2-2}$$

where the names for the angles and sides are shown in Fig. 2-3(a). Both of these expressions may be written in many different forms, depending on which angle or side is unknown. You should also recognize that the cosine law becomes the Pythagorean theorem when γ is a right angle, as shown in Fig. 2-3(b). In that case, cosine $90° = 0$, and the equation becomes $C^2 = A^2 + B^2$.

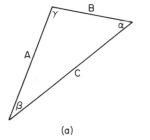

(a)

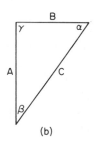

(b)   **Figure 2-3**

2-3. ACCURACY AND SIGNIFICANT FIGURES

For most of our work, the use of three significant figures, as can be obtained on a 10-in. slide rule, gives sufficient accuracy; the use of three significant figures implies an accuracy of between 0.1 and 1.0 percent.

In reality, the loads on structures will frequently not be known with that high a degree of accuracy, for the loads will be based on a reasonable estimate. What is a reasonable estimate of the maximum load on a chair, a crane, or a bridge? We do not know for certain. However, we will treat the loads indicated in problems in this book as being accurate to three significant figures.

In certain problems where we need to subtract numbers, it may be necessary to use more than three significant figures in order to arrive at three significant figures in the answer. For example, $40.1 - 39.8 = 0.3$ gives an answer with only one significant figure. However, if we had more accurate information using numbers with five significant figures, there would then be three significant figures in the answer, as in $40.126 - 39.783 = 0.343$.

If, in your solution, you obtain an answer with more significant figures than you have in the given information, then you should round off your answer to the same number of significant figures contained in the given information with the fewest significant figures. For example, in work with three significant figures, 9.8746 would be rounded to 9.87, and 0.036551 would be rounded to 0.0366.

2-4. UNITS OF MEASUREMENT

The British system of units will be used throughout this book, since they are commonly used in engineering in English-speaking countries. The length units most frequently used are inches and feet. The pound and kilopound (usually referred to as the *kip*, which equals 1000 pounds) are the units of weight most commonly used in applied mechanics. The second and minute are the most commonly used units of time. Other units of measure will also be used occasionally but not as often as those listed above.

2-5. GRAPHICAL METHOD

Many of the problems of mechanics can be solved using graphical methods. There are two types of solutions which are graphical. In one type, a graph is drawn, showing how one quantity varies with another (e.g., speed might vary with time). From such a graph it is often possible to calculate other information which we need. This type of solution will be discussed in detail in Chapter 8.

The other type of graphical solution involves using some drafting skills to draw triangles and polygons where the lengths of the sides are related to physical quantities such as force, displacement, or velocity. For graphical solutions you will need some of your drafting equipment: hard, sharp pencils, triangles, an engineer's scale, a compass, and a protractor. Neatness and accuracy will be important in your solutions, just as they are in any drafting assignment.

You can begin using graphical methods of solution to solve problems in trigonometry by drawing the triangles to scale, based on the information given in the problems at the end of the chapter. You will find that although all the information about a triangle is not given, there is enough information for you to draw the triangle if you use a bit of ingenuity.

2-6. EXAMPLE PROBLEMS

Following most of the topics in this book you will find one or more example problems illustrating the principles which have just been presented. Where possible, both a graphical and analytical solution will be shown. The comments on the left-hand side of the example–problem solutions are our thoughts as we go about solving the problems. The actual solution, as it should be worked out, is on the right-hand side of the page.

EXAMPLE 2-1

A triangle has side $A = 2$ in., $C = 3$ in., and an angle of $\beta = 130°$ between the two given sides. Find the third side, B, and the other two angles, α and γ, which are opposite sides A and C, respectively.

The information given is two sides and the contained angle. By using the cosine law, the third side can be found. The cosine of $130°$ is equal to $-\cos(180° - 130°) = -\cos 50°$.

$$B^2 = A^2 + C^2 - 2AC \cos \beta$$
$$= 2^2 + 3^2 - 2 \times 2 \times 3 \times (-.643)$$
$$= 4 + 9 + 7.71$$
$$= 20.71$$
$$B = 4.55 \text{ in.}$$

As all three sides are now known, the sine law can be used to find α.

$$\frac{\sin \alpha}{A} = \frac{\sin \beta}{B}$$
$$\sin \alpha = \frac{2 \times .765}{4.55} = .336$$
$$\alpha = 19.6°$$

Since the sum of the interior angles in a triangle is $180°$, this fact can be used to find γ.

$$\gamma = 180° - \alpha - \beta$$
$$= 180° - 19.6° - 130°$$
$$= 30.4°$$

To check the solution graphically, draw line $C = 3$ in. and draw line $A = 2$ in. at an angle of $130°$ with line C as shown in Fig. 2-4. The line closing the triangle must be B. Its length can be measured as can angles α and γ.

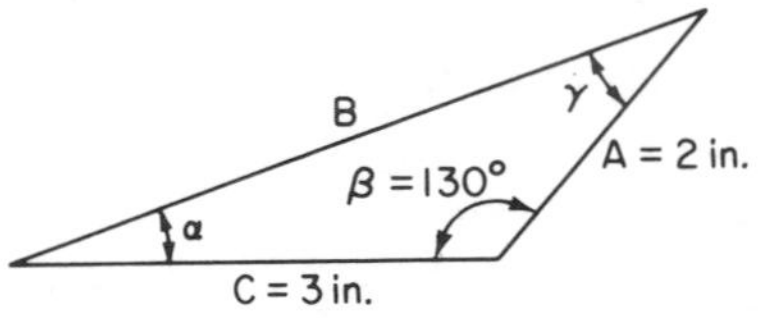

Figure 2-4

PROBLEMS

2-1. For the four angles $15°$, $40°$, $85°$, and $120°$, find the value of each of the six trig functions by (a) looking up the values in a set of math tables, (b) finding the values on your slide rule, and (c) drawing right-angled triangles with one of the angles equal to the given angle and measuring the lengths of the sides to obtain the ratios for the trig functions.

2-2. The shortest side of a right-angled triangle is 1.50 in., and the angle opposite the shortest side is $40°$. Determine (a) the length of the hypotenuse, (b) the length of the third side, and (c) the third angle.

2-3. A right-angled triangle has sides of 12 in. and 16 in. Calculate (a) the hypotenuse, (b) the sine, (c) the cosine, (d) the tangent, and (e) the ratio $\sin \theta / \cos \theta$ for the smallest angle.

2-4. For a triangle with side $A = 6$ in., side $B = 8$ in., and the angle $\alpha = 28°$ opposite A, find the length of the third side C and the values for the other two angles, β and γ. There are two possible solutions for this problem.

2-5. A triangle has one angle of $35°$ and a second angle of $65°$. The side opposite the $35°$ angle is 2.65 ft. Determine the size of the third angle and the lengths of the other two sides.

2-6. A triangle has sides of 6, 7, and 8 in. Determine the size of each of the angles in the triangle.

2-7. Two parts of a frame on a machine are 8 in. and 4 in. long, respectively. A brace is required to support the two parts. If the angle between the parts is 30°, determine the length of the brace required at the end of the parts and the angles the brace makes with the other two parts of the frame.

2-8. Find the length of the third side and the sizes of the other two angles for the triangle with sides of 5 ft and 9 ft, and a contained angle of 115°.

2-9. A cam mechanism has the dimensions shown in Fig. P2-9. Determine the distance

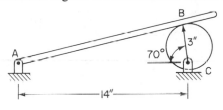

Figure P2-9

from the pivot point of the follower *A* to the contact point *B*.

2-10. A surveyor is to determine the distance between two points *A* and *B*, on opposite sides of a river. To do this, he sets his transit at *B*, and locates a third point *C*. He mearures the angle at *B*, and then moves his transit to *C*, where he measures the angle at *C*, and the distance from *B* to *C*, as shown in Fig. P2-10. Determine the distance between points *A* and *B*.

2-11. A vacant lot is a rectangle 50 ft by 150 ft. How much shorter is the distance walked by cutting along the diagonal instead of walking along the two edges?

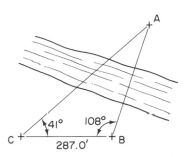

Figure P2-10

2-12. A 400-ft radio antenna is to be guyed. On one side the land slopes up at 1 ft to 8 ft from the base of the antenna as shown in Fig. P2-12. If the angle between the guy and the antenna must be 35°, how long must the guy be, and what distance is measured along the ground from the base of the antenna to the guy anchor point to locate the anchor?

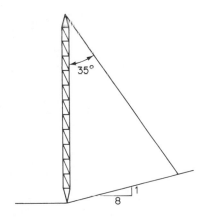

Figure P2-12

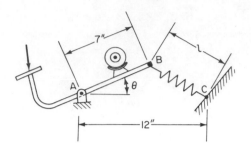

Figure P2-13

2-13. Figure P2-13 shows part of a brake mechanism. If the length, l, of the spring is stretched greater than 8 in., the spring will fail. Determine the maximum possible value for the angle, θ.

3

FORCES, VECTORS, AND RESULTANTS

We will spend a great deal of our time studying forces and their effects on bodies, so it is logical to begin the actual study of applied mechanics with an examination of what forces are and the way in which they behave.

3-1. FORCE DEFINED

A force can be defined as the action of one body on another. There are several ways that a body can exert a force on another body. The simplest of these methods is by contact between bodies. If you set this book on the desk, the book exerts a force on the desk through contact, and similarly, the desk exerts an upward force on the book. The second type of force, and perhaps the most common, is the gravitational force. This occurs because of the mutual attraction between any two bodies. Again, if you take this book, hold it above your desk, and then let go of it, even though the book is not touching anything, there is still a force exerted on it by the earth's gravitational pull. The book is also pulling on the earth with the same force that the earth pulls on the book. However, this force does not affect the motion of the earth to any noticeable extent. Gravitational attraction occurs between

17

Fig. 3-1. Sailing is a sport where a working knowledge of mechanics is necessary for success. Most sailors have not had formal training in applied mechanics, but have learned through experience the concepts of adding vectors and finding resultants in order to get the best performance from their boats. (Photo courtesy of the International Lightning Class Association.)

any two bodies. When neither of the bodies is large, the size of the gravitational force is so small that it is usually neglected. A third type of force, which you will not encounter very often, is the magnetic force, which exists because of the magnetic attraction or repulsion that exists between two magnets, or between a magnet and some metals.

If you consider the forces mentioned above, you will recognize that certain requirements must be met in describing a force in order to determine the effect of that force on a body. First, we must know the *magnitude*, or size, of the force, which tells us how big the force is. Secondly, we need to know the *direction*, or line of action, in order to determine the orientation of the force. Included in the direction of the force must be the *sense* which indicates whether we are dealing with a push or a pull. Finally, to fully describe the force, we need to know the *point of application*. This will have a great influence on the behavior of a body under the influence of the force.

A recreational application of an understanding of the behavior of forces is shown in Fig. 3-1. In fact, some sailors take their sailing so seriously that students of applied mechanics may find that if they have a sailing acquaintance they can get some extra help from him in their studies.

Figure 3-2 shows a force applied to a body. The force is represented by the arrow, with the length of the arrow in proportion to the magnitude of the force to some convenient scale. The direction is measured relative to some convenient reference; in this case it is the angle θ above the horizontal axis. The sense is the direction in which the arrow points, in this case down to the left. Finally, the force is shown as passing through a particular point, P, on the body.

To further illustrate the importance of using all the information required to describe a force, consider the small car shown in Fig. 3-3. Assume that in each case a force of the same magnitude or size is exerted on the car. In (a) the car would move to the left, in (b) it would move to the right, in

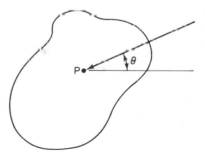

Figure 3-2

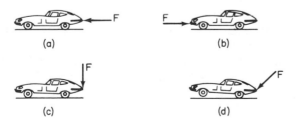

(a)   (b)

(c)   (d)   **Figure 3-3**

(c) the back would move down, and the front would move up, and in (d) the back of the car would move down, and the car would tend to move to the left. Each of the four forces shown is not the same. Although they are the same size, it is fairly apparent that each would have a different effect on the action of the car. Because of the different effects of different forces, it is very important to fully describe each force by including a description of the magnitude, direction, sense, and point of application.

A quantity which requires a magnitude and direction to describe it is a *vector quantity*. A force is one of many quantities which are vectors.

3-2. VECTORS AND SCALARS

A *scalar* is a quantity which can be described in terms of size or magnitude only. Examples of scalar quantities are age, length, area, and mass. Some examples of vectors, which must have magnitude and direction, are force, displacement, velocity, and acceleration.

To help differentiate between vectors and scalars in this book, all vectors will be shown in boldface type: **A**. Italic Lightface type, A, will be used to indicate the magnitude of a vector. Scalar quantities are also indicated by italics. Since you cannot write in boldface, the standard practice for written work is to put an arrow over the vector quantities, as shown $\vec{A}$. Generally, upper case or capital letters are used for vectors, although there is no rule to this effect, and there are many exceptions.

You have been using scalars for many years in your mathematics and are already familiar with addition, subtraction, multiplication, and division with scalars. Since vectors differ from scalars, you might expect that vectors will behave differently under many circumstances, and they do. The addition, subtraction, and multiplication of vectors requires the use of vector mathematics, which is somewhat different from what you are accustomed to. At the moment, there is a need to be concerned only with vector addition and subtraction. The methods are not difficult, as you will find later on in this chapter. Just remember that, generally, the sum of two vectors is different than the sum of two scalars.

3-3. VECTOR ADDITION

The sum of two or more vectors is called the *resultant*. There are several rules for obtaining vector sums, all of which involve, either directly or indirectly, a graphical procedure. These rules are perhaps best illustrated by referring to Fig. 3-4, which shows two points on the ground, *a* and *b*. The straight-line path, *ab*, is a displacement. This is a vector, for it has a magnitude and a direction. It also has a sense and a starting point, which is a point of application. Parts (b) and (c) of Fig. 3-4 also show vectors **C** and **D**, and **E** and **F**. If you started at *a* and followed displacement **C** and then followed displacement **D**, you would end up at point *b*. Hence, the displacement **C** followed by displacement **D** gives exactly the same displacement as **R**. Therefore, the vector sum of **C** and **D** is equal to **R**. This would be written as **C** + **D** = **R**. In a similar fashion, we also have that **E** + **F** = **R**. This is an illustration of the *triangle rule* for vector addition, which states that two vectors may be added by drawing them tip to tail. Their sum or resultant is the vector from the tail of the first vector to the tip of the second vector.

Two vectors may also be added using the *parallelogram rule*, in which the two vectors are drawn starting from a common point, as shown in Fig. 3-5. A parallelogram is drawn, using the two vectors as two adjacent sides. The diagonal of the parallelogram from the origin of the two vectors is the resultant, or sum, of the two vectors. You will note from the parallelogram that it is really made up of two triangles, and that the parallelogram really looks like two vector triangles solving the vector equations **A** + (**B**) = **R** and **B** + (**A**) = **R**. The parallelogram law then helps to illustrate another feature of vectors: the sequence or order of addition does not affect the sum of the vectors.

The most general rule for vector addition is the *polygon rule*. It is illustrated in Fig. 3-6. According to the polygon rule, the resultant of several vectors may be obtained by adding the vectors together tip to tail to form a polygon, called a vector polygon. The sum or resultant is that vector from the origin of the polygon to the tip of the last vector. It is written as **A** + **B** + **C** + **D** = **R**. Actually, the triangle rule is just a special case of the polygon rule, since a polygon with three sides is a triangle.

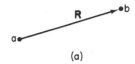

(a)

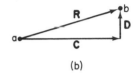

(b)

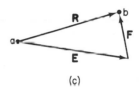

(c)

Figure 3-4

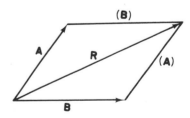

Figure 3-5

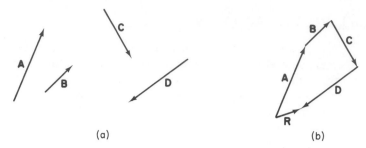

Figure 3-6   (a)   (b)

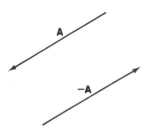

Figure 3-7

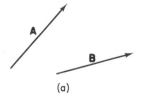

(a)

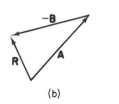

(b)

Figure 3-8

Up to this point, we have discussed vector addition in terms of solving the problems graphically. Before considering an analytical approach to solving our problems, let us consider subtracting vectors. First we must define a negative vector. Referring to Fig. 3-7, we see vector **A**. Vector $-$**A** is the vector which is equal and opposite to vector **A**. If we accept this definition, then vector subtraction is no different than vector addition.

Suppose we are given the two vectors **A** and **B** shown in Fig. 3-8(a), and we are to find **R** where **R** = **A** $-$ **B**. This equation is the same as **R** = **A** + ($-$**B**). The resultant is shown in Fig. 3-8(b).

In order to solve vector addition problems analytically, in most cases it is necessary to make a sketch. The method of solution is to solve for the resultant as the unknown side of a triangle, using either the sine law or the cosine law, depending on the information we have about the triangle. This statement implies that we can solve only vector addition (or subtraction) problems where two vectors have been added together tip to tail as in the triangle law for addition. This is basically true, for if more than two vectors are to be added together, we would need to pair two vectors, add their resultant to the next vector, then add this resultant to the next vector, and continue with the process until all vectors have been added. Since this process would be a tedious method of adding many vectors, another method of adding more than two vectors analytically will be discussed following the example problem.

EXAMPLE 3-1

Two vectors are shown in Fig. 3-9(a). Find (a) **A** + **B** and (b) **A** $-$ **B**. Assume that **A** has a magnitude of 45 units and **B** has a magnitude of 60 units.

Since a sketch will be very helpful in the solution, the problem might as well be solved graphically first. The vectors are shown added together tip to tail in Fig. 3-9(b). The resultant is the vector which closes the triangle.

The vector difference is obtained by reversing the direction of **B** and adding to the tip of **A** as shown in Fig. 3-9(c).

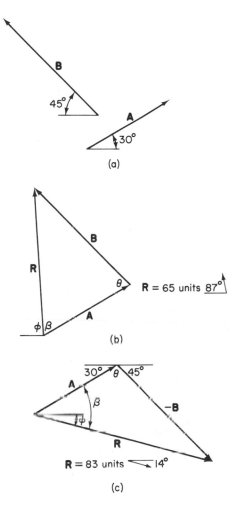

(a)

(b)

(c)

Figure 3-9

From the direction of the vectors θ can be found. With two sides and the contained angle known, the cosine law is used to find the third side.

(a) $\theta = 45° + 30° = 75°$

$$R^2 = A^2 + B^2 - 2AB \cos \theta$$
$$= 45^2 + 60^2 - 2 \times 45 \times 60 \times .259$$
$$= 2025 + 3600 - 1400$$
$$= 4225$$
$$R = 65.0 \text{ units}$$

It is necessary to find the direction of **R**. If angle β is found then the direction of **R** can be found, since the direction of **A** is already known.

$$\frac{\sin \beta}{B} = \frac{\sin \theta}{R}$$

$$\sin \beta = \frac{60 \times .966}{65} = .891$$

$$\beta = 63°$$
$$\phi = 180° - 30° - 63°$$
$$= 87°$$

The reasoning for (b) is the same as for (a).

(b) $\theta = 180° - 30° - 45° = 105°$
$$R^2 = A^2 + B^2 - 2AB \cos \theta$$
$$= 45^2 + 60^2 - 2 \times 45 \times 60$$
$$\times (-.259)$$
$$= 2025 + 3600 + 1400$$
$$= 7025$$
$$R = 83.9 \text{ units}$$
$$\frac{\sin \beta}{B} = \frac{\sin \theta}{R}$$
$$\sin \beta = \frac{60 \times .966}{83.9} = .691$$
$$\beta = 43.8°$$
$$\phi = 43.8° - 30° = 13.8°$$

3-4. VECTOR COMPONENTS

The reverse of adding vectors is breaking them up into components, called *resolution of vectors*. Components are vectors, and the sum of all the components of a vector must equal the original vector. Three examples are shown in Fig. 3-10. In each case the vectors **B** and **C** are the components of vector **A**. Conversely, in each case, **A** is the resultant of **B** and **C**. Components can be obtained either graphically or analytically. The procedures will be the same as for finding resultants, except that generally when seeking components you will be given two angles and one side, whereas when finding resultants you were given two sides and one angle as known quantities in your triangles. Special attention is to be given to the components of Fig. 3-10(b) and 3-10(c); these two figures show rectangular components. These are components at right angles to each other and are often very useful in solving problems. Frequently we think of rectangular components in terms of vertical and horizontal components, or *x* and *y* components. However, as illustrated in Fig. 3-10(c), there is no need for rectangular components to be vertical or horizontal.

Finding rectangular components is usually very simple because we are dealing with a right-angled triangle. Conse-

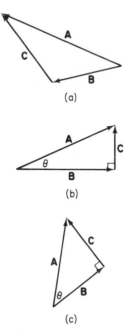

(a)

(b)

(c)

Figure 3-10

quently, as shown in Fig. 3-10(b) and 3-10(c), the magnitude of the components are $C = A \sin \theta$ and $B = A \cos \theta$.

When dealing with rectangular components, we frequently write them with subscripts indicating the axis to which components are parallel, e.g., A_x, A_y, F_x, and F_y are the components of **A** and **F** parallel to the x and y axes, respectively.

To indicate direction, signs are used. An x component pointing to the right is positive, one pointing to the left is negative. A y component pointing up is positive and one pointing down is negative. Thus the expression $F_x = -4\,\mathrm{lb}$ would indicate a 4-lb force parallel to the x axis and pointing to the left.

When a force is being replaced by its components, in general the components must intersect on the line of action of the original force as shown in Fig. 3-11. If the components

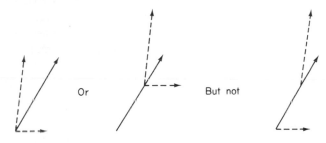

Or   But not   Figure 3-11

do not intersect on the line of action of the original force, the components will not be completely equivalent to the initial force. The reason for this will be apparent after you read Section 3-10 on nonconcurrent force systems later in this chapter. However, we may still use the vector triangle as an aid to solving for the magnitude of the components.

3-5. VECTOR ADDITION USING COMPONENTS

If we have several vectors to add together, they may be added graphically using the vector polygon. To add them analytically we sum components in the same direction. In other words, all the x components of the vectors are added together, all the y components are added together, and the resultant is obtained. This can be stated mathematically as:

$$R_x = \Sigma F_x \qquad (3\text{-}1)$$
$$R_y = \Sigma F_y \qquad (3\text{-}2)$$
$$R = (R_x^2 + R_y^2)^{1/2} \qquad (3\text{-}3)$$

The angle of the resultant can be obtained by using R_x and R_y to find the tangent of the angle between the horizontal and the resultant, where

$$\tan \theta = \frac{R_y}{R_x} \qquad (3\text{-}4)$$

(a)

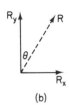

(b)

Figure 3-12

Generally the angle designated as θ is that angle less than $45°$, since the values for $\tan \theta$ are larger than 1.0 if θ is larger than $45°$; these values cannot be found directly on many slide rules. Thus in some cases, $\tan \theta = R_x/R_y$ will also be used. In solving these problems, it is very important to make a small sketch of R_x and R_y as shown in Fig. 3-12 in order to clarify which angle is θ.

It should be noted that, when vectors parallel to the same axis are added, the magnitude of the resultant of the vectors is simply the algebraic sum of the vectors. It must be emphasized that only when the vectors are parallel to the same axis can they be added algebraically.

EXAMPLE 3-2

Find the vertical and horizontal components of the force shown in Fig. 3-13(a).

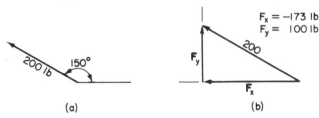

(a)   (b)

Figure 3-13

Perhaps an analytical solution will be simpler if we first draw a sketch as shown in Fig. 3-13(b). We can use this sketch for our graphical solution if it is drawn to scale. A horizontal line is drawn through the tail of the vector and a vertical line is drawn through the tip. Since $\mathbf{F}_x + \mathbf{F}_y = \mathbf{F}$, then the point of intersection of the vertical and horizontal lines will represent the tip of $\mathbf{F}_x$ and the tail of $\mathbf{F}_y$, as shown in the figure. The values can be scaled.

From the triangle of our sketch, we can calculate F_x and F_y. The signs depend on the direction in which F_x and F_y point.

$$F_x = -200 \cos 30°$$
$$= -173.3 \text{ lb}$$
$$F_y = 200 \sin 30°$$
$$= 100 \text{ lb}$$

If you prefer, you may use the 150° angle and obtain the proper sign for the component from the trig function.

$$F_x = 200 \cos 150°$$
$$= -173.3 \text{ lb}$$
$$F_y = 200 \sin 150°$$
$$= 100 \text{ lb}$$

EXAMPLE 3-3

For the force shown in Fig. 3-14(a), find the component along the m and n axes.

A sketch will help in solving the problem, so the graphical solution might as well be done first. A line parallel to the n axis is drawn from the tip of the vector to intersect the m axis. This point of intersection is the tip of the vector along the m axis and the tail of the vector parallel to the n axis. The necessary construction is shown in Fig. 3-14(b).

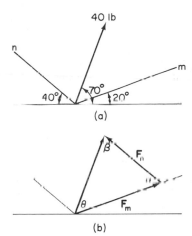

(a)

(b)

Figure 3-14

From the geometry of the construction we can determine θ, β, and α. With one side and all angles known, the sine law can be used to find the other sides.

$$\theta = 70° - 20° = 50°$$
$$\beta = 180° - 70° - 40° = 70°$$
$$\alpha = 180° - 50° - 70° = 60°$$

$$\frac{F_m}{\sin \beta} = \frac{F}{\sin \alpha}$$

$$F_m = \frac{40 \sin 70°}{\sin 60°} = 43.4 \text{ lb}$$

$$\frac{F_n}{\sin \theta} = \frac{F}{\sin \alpha}$$

$$F_n = \frac{40 \sin 50°}{\sin 60°} = 35.4 \text{ lb}$$

EXAMPLE 3-4

Figure 3-15 shows three vectors **A**, **B**, and **C** which have lengths of 7, 6, and 4 units, respectively. Find the resultant of the three vectors.

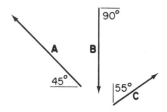

Figure 3-15

To solve analytically, first find the sum of the x and y components. Signs depend on whether the components are headed in the positive or negative x or y direction. Whether the sine or cosine is used to find a horizontal component depends on whether the angle is measured from the vertical or horizontal.

$$\Sigma F_x = -7 \cos 45° + 6 \cos 90° + 4 \sin 55°$$
$$= -4.949 + 0 + 3.272$$
$$= -1.673 \text{ units}$$
$$\Sigma F_y = 7 \sin 45° - 6 \sin 90° + 4 \cos 55°$$
$$= 4.949 - 6.00 + 2.296$$
$$= 1.245 \text{ units}$$

The resultant of the two rectangular components is obtained by the Pythagorean theorem. Indicating the orientation of the resultant is an important part of the solution. Note that in this problem either R or θ may be found first.

$$R = (R_x^2 + R_y^2)^{1/2}$$
$$= [(-1.673)^2 + (1.245)^2]^{1/2}$$
$$= (2.80 + 1.55)^{1/2}$$
$$= (4.35)^{1/2}$$
$$= 2.08 \text{ units}$$

$$\tan \theta = \frac{1.245}{1.673} = .744$$

$$\theta = 36.6°$$

Graphically, the vectors are drawn tip to tail to scale, as shown in Fig. 3-16. The resultant is the vector from the origin of the first vector to the tip of the last vector. The length and direction can be obtained with scale and protractor.

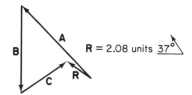

Figure 3-16

3-6. RESULTANTS USING THE SLIDE RULE

If the two components whose resultant is desired are perpendicular (as is frequently the case when components have been obtained), the resultant and its direction can be obtained quickly and accurately on most slide rules without having to use the Pythagorean theorem. The procedure is as follows:

1. Set the index of the C scale over the larger of the components on the D scale.
2. Set the hairline on the cursor over the smaller component on the D scale.
3. The angle between the large component and the resultant may now be read on the T scale under the hairline.
4. Without moving the cursor, set the same angle but on the S scale under the hairline.
5. The resultant can now be read on the D scale under the index of the C scale.

Try this procedure for a couple of simple cases, for example, when the components are 3 and 4 units (the resultant is 5) and when the components are 5 and 12 units (the resultant is 13).

Notice that this procedure will not work if the ratio of the larger number to the smaller number is greater than 10 to 1. In this case the resultant is the same as the larger number for three significant figures, and the value of the angle may be obtained using the ST scale on the slide rule.

PROBLEMS

3-1. A vector 6 units long in the *x* direction is added to a vector 4 units long in the −*y* direction. Find the resultant.

3-2. Find the resultant for the two vectors shown in Fig. P3-2.

3-3. A man walks south 200 ft and then east for 150 ft. Determine the direction and distance he would have traveled if he had taken a direct route from the starting point to the end point.

3-4. For the vectors **C** and **D** shown in Fig. P3-4, find **C** + **D** and **C** − **D**.

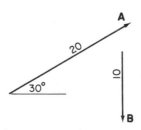

Figure P3-2

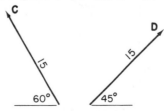

Figure P3-4

3-5. Vector **E** is 5 units long at a positive angle of 50° with the horizontal and vector **F** is horizontal, 7 units long, and in the negative *x* direction. Find **E** + **F** and **E** − **F**.

3-6. A man walks 2 miles in a direction 35° north of east. Determine the east and north components of this displacement.

3-7. Determine the vertical and horizontal components of a vector 12 units long which makes a clockwise angle of 40° with the vertical.

3-8. A force of 500 lb is directed horizontally to the right. Divide this force into rectangular components, with one component at 30° below the horizontal.

3-9. An airplane flies 60 miles on a course 60° south of west. What distance in the due south and due west directions would it have traveled if the trip had been made following these two component directions?

3-10. A 300-lb force points vertically upward. One component of this force is 450 lb and is directed 25° clockwise from the 300-lb force. Determine the magnitude and direction of the other component.

3-11. Determine the *x* and *y* components for the force shown in Fig. P3-11 for (a) $\theta = 25°$ and (b) $\theta = 50°$.

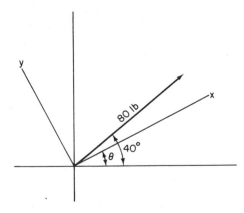

Figure P3-11

3-12. Find the resultant of the three given vectors. **A** is 40 units at 15°, **B** is 25 units at 100° and **C** is 15 units at 240°. All angles are positive and measured from the positive *x* axis.

3-13. For the vectors shown in Fig. P3-13,

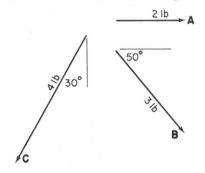

Figure P3-13

find (a) **A** + **B** + **C**, (b) **A** − **B** − **C**, and (c) **A** − **B** + **C**.

3-14. After being hit, a baseball has velocity components of $v_x = 60$ ft/sec and $v_y = 25$ ft/sec. Keeping in mind that velocities are vectors and must be added vectorially, what is the total velocity and direction of the baseball?

3-15. A three-jawed chuck, as used on a metalworking lathe to hold the part being machined, exerts three forces on the part as shown in Fig. P3-15. If each of the forces is 190 lb, determine the resultant of the three forces.

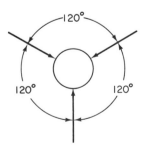

Figure P3-15

3-7. RESULTANTS OF CONCURRENT FORCE SYSTEMS

Now that we have developed an ability to find resultants, we can use this ability to find the resultant of plane, concurrent force systems. First, of course, we must define our terms. A *plane force system* is a group of forces all in one plane. *Concurrent forces* are forces which have a common point of intersection. If we assume that the forces shown in Fig. 3-17 are in a plane, then the three forces **A**, **B**, and **C** constitute a plane, concurrent force system. At first glance, **A**, **B**, and **C** do not appear to be concurrent. However, the dashed-line projections of **A** and **C** do meet at a common point *P*. This point *P* is the point of concurrency. We may move forces along their lines of action (providing they are acting on a rigid body) without changing their effect on the body. This is called the *principle of transmissibility*.

The resultant of any such plane, concurrent force system can be obtained using any of our methods of finding the resultant of vectors. This resultant force must be applied along a line which passes through the point of concurrency of the initial force system. If its line of action does not pass through the point of concurrency, the resultant will not have the same effect on the body as the original system of forces.

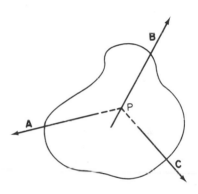

Figure 3-17

PROBLEMS

3-16. Two forces **A** and **B** intersect at a point. If **A** is 75 lb at 25° below the horizontal pointing to the right and **B** is 30 lb at 30° above the horizontal pointing to the right, find the resultant of the two forces.

3-17. Two water skiers are being pulled by a boat as shown in Fig. P3-17. If the force in tow rope *A* is 40 lb and the force in tow rope *B* is 65 lb, find the resultant force that the two ropes exert on the boat.

3-18. Replace the three forces shown in Fig. P3-18 by a single force. (See figure.)

3-19. Three forces are applied to an eye as shown in Fig. P3-19. What is the equivalent single force that the eye must resist? (See figure.)

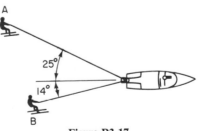

Figure P3-17

3-20. Find the resultant of the four forces shown in Fig. P3-20. (See figure.)

3-21. Determine the resultant of the four forces shown in Fig. P3-21. (See figure.)

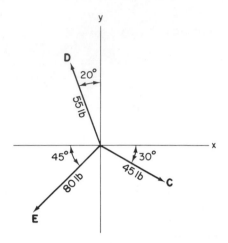

Figure P3-18

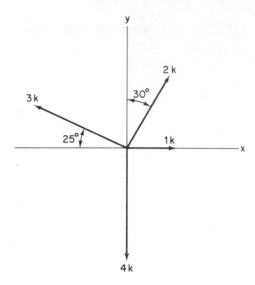

Figure P3-20

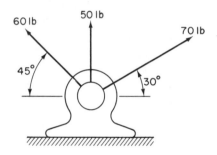

Figure P3-19

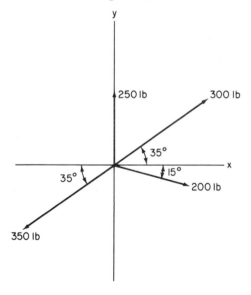

Figure P3-21

3-8. MOMENT OF A FORCE

Sometimes a force is defined as a push or a pull. You would expect that when you push or pull on a body it will travel in a straight line along the direction of the push or pull.

However, you have probably observed that a body does not always move in such a manner. If you leaned against the top shelf of a tall bookcase, the bookcase is more likely to tip over or rotate than it is to slide. Similarly, if you push horizontally on one corner of your desk along the short side, the desk is most likely to turn or rotate rather than move in a straight line. This rotation is due to the *moment of the force.* Moment is defined as the tendency to cause rotation about some axis. By experimenting with your desk or a book, you will conclude that the tendency to rotate depends on the size of the force and its point of application.

The moment of a force about an axis is the product of the force and the perpendicular distance between the line of action of the force and the axis. It is expressed as

$$M = Fd \qquad (3\text{-}5)$$

where

M is the moment,

F is the force causing the moment, and

d is the perpendicular distance from the axis about which moments are taken to the line of action of the force.

The axis about which the moment is taken is frequently called the axis of rotation, and the perpendicular distance d is frequently called the moment arm. The axis of rotation shows as a point on a two-dimensional drawing. Hence, moments are often said to be taken about a point. If the force is in pounds and the distance is in feet, the units of moment will be foot-pounds (ft-lb). Other units which will be used are inch-pounds (in-lb) and foot-kips (ft-k).

A moment is a vector quantity. The line of action of the moment vector passes through the point about which moments are taken and is perpendicular to the plane formed by the force and the line from the point to the force. The direction is obtained by the *right-hand rule.* According to the right-hand rule, if we curl the fingers of the right hand in the direction of rotation, the thumb will point in the direction that the vector should point. Figure 3-18(a) shows the force **F**, the perpendicular distance or moment arm d to the point P, and the moment **M**. The curved arrow shows the curl of the fingers used to obtain the direction of **M**.

There is a common sign convention for moments. If the moment vector points out of the page, it is positive, and if it points into the page, it is negative. It is sometimes stated that clockwise rotation is negative, and counterclockwise

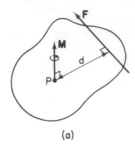
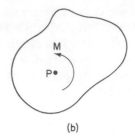

Figure 3-18　　　　　(a)　　　　　　　　　　　　　　　　(b)

rotation is positive. The sign of the moment does not depend on the sign of the force or the sign of the direction, but only on the direction of rotation which would be caused.

Drawing good three-dimensional figures requires a great deal of skill, so we seldom draw figures like Fig. 3-18(a). Instead, the moment is usually represented by a curved line in the same plane as the force and distance as shown in Fig. 3-18(b) which shows how the moment **M** about the axis of rotation *P* due to the force **F** would be indicated.

In some types of problems, finding the perpendicular distance from a force to the point about which moments are to be taken is a time-consuming operation. It is frequently possible to simplify the task by using *Varignon's theorem*, which states that the sum of the moments of the components of a force about a point is the same as the moment of the force about that same point. The application of this principle is illustrated in the following example problem. Note that if the moment vectors are parallel, as they will be in a plane problem, the resultant will be the algebraic sum of the moments.

EXAMPLE 3-5

Find the moment about point *P* of the force shown in Fig. 3-19.

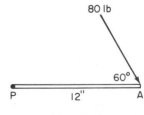

Figure 3-19

First the perpendicular distance from the force to the point must be calculated. The

$d = 12 \sin 60°$
$\quad = 10.4 \text{ in.}$

distance is shown in Fig. 3-20(a). The moment can then be calculated.

$M_P = Fd$

$= -(80 \times 10.4)$

$= -832$ in-lb

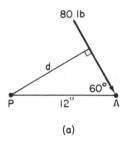

(a)

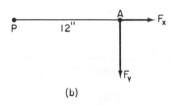

(b)

Figure 3-20

Alternately, to use Varignon's theorem, first determine if the use of any particular components will simplify the problem. If horizontal and vertical components concurrent at A are used as shown in Fig. 3-20(b), the moment of the horizontal component about P will be zero, and only the moment of the vertical component will need to be calculated.

$F_y = -(80 \sin 60°)$

$= -69.3$ lb

$M_P = F_y d$

$= -(69.3 \times 12)$

$= -832$ in-lb

PROBLEMS

3-22. For the force shown in Fig. P3-22, determine the moment about (a) point A and (b) point B.

3-23. A wrench is shown in Fig. P3-23. Determine the moment of the 40-lb force about the bolt. What is the moment about the bolt if the 40-lb force was applied perpendicular to the wrench at the same point of application? (See figure.)

3-24. Find the sum of the moments of the 8- and 10-kip forces about point C, as shown in Fig. P3-24. (See figure.)

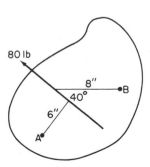

Figure P3-22

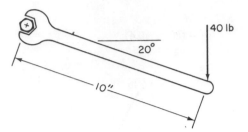

Figure P3-23

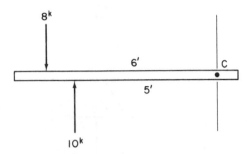

Figure P3-24

3-25. A pulley driven by a belt is shown in Fig. P3-25. If the tension in part *A* of the belt is 20 lb and the tension in part *B* is 84 lb, determine the moment the belt causes about point *C* at the center of the pulley.

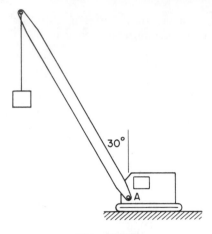

Figure P3-26

3-26. A crane with a 90-ft boom picks up a 5-ton weight, as shown in Fig. P3-26. What is the moment of the weight in ft-lb about point *A* at the front of the crane?

3-27. Find the sum of the moments of the 60-lb force **F** and the 45-lb force **G** about point *A*, as shown in Fig. P3-27.

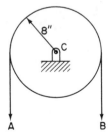

Figure P3-25

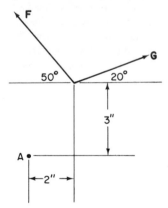

Figure P3-27

3-9. COUPLES

In addition to forces and moments, there is another force-related quantity with which you must become familiar: a *couple*. A couple is a pure moment and may be considered to be the resultant of two parallel forces of equal magnitude and opposite direction. Such a system of forces is shown in Fig. 3-21. You can see that the vector sum of the forces is zero. However, each of these forces tends to cause rotation about some point. The sum of the moments about point E on $-\mathbf{A}$ will be Ad. Similarly the sum of the moments about point D on $\mathbf{A}$ will be Ad. In both of the above cases, the moments of the two forces about a point have the same value. If moments are summed about some random point G which is not on either of the forces, the value for the moment is

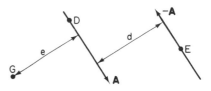

Figure 3-21

$$M_G = -(Ae) + [A(d + e)]$$
$$= -(Ae) + Ad + Ae$$
$$= Ad$$

In other words, if we take the moments of these two forces about any point, we will find that the sum of the moments is Ad. The couple represented by the two forces would then be expressed as

$$C = Ad$$

In general, any system of forces which has a resultant force of zero and which causes a moment about some point can be replaced by a couple.

One of the basic differences between moments and couples is that the size of the moment depends on the location of the point about which the moment is taken, whereas a couple will have the same magnitude regardless of the reference point for the couple. Unfortunately, in much of the technical literature, the terms are incorrectly used. For instance, in problems on stresses in beams, the term "moment" is generally used, when in reality it would be correct to use the term "couple". Since the error is well established, we will live with it rather than try to force a change. However, in this book the symbols C and M will be used to differentiate between the couple and the moment.

Couples are vectors whose directions are also obtained using the right-hand rule. The couple vector is a *floating vector* in that it has no particular point of application, while the moment vector must pass through the point about which the moments were taken. Vector addition is used with couples, but wherever the couples are parallel the couples may be added algebraically. The sign convention for couples is the same as for moments; clockwise rotation is negative and counterclockwise rotation is positive.

You may find the term *torque* used instead of couple. Basically the two are the same, but common usage is for torque to be a couple transmitted by a shaft, such as in an automobile transmission, while a couple is usually considered to be the effect of a pair of equal and opposite forces.

There may appear to have been an inconsistency in notation in the discussion about moments and couples. Moments and couples are vectors, but boldface type has not been used for all the moments and couples. The reason is that since all problems to date have been plane problems, the moment or couple vectors have always been perpendicular to the plane, so that the use of the + or − sign has been sufficient to indicate all information about the vector when used with the italic lightface type indicating magnitude.

The vector mathematics required to write moment or couple equations in vector form is a complication that we can manage to do without when most of our problems are plane problems.

EXAMPLE 3-6

Find the size of the couple shown in Fig. 3-22.

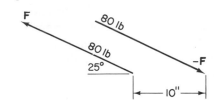

For finding the magnitude of a couple, it is best if moments are taken about a point on one of the forces, since the moment of that force about the point will be zero.

Figure 3-22

The perpendicular distance d between the forces is shown in Fig. 3-23 and can be calculated from the right triangle formed.

$d = 10 \sin 25°$

$\quad = 4.23$ in.

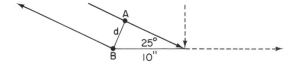

Figure 3-23

The couple may then be obtained by finding the moment of **F** about *A*.

$$C = Fd$$
$$= -(80 \times 4.23)$$
$$= -338 \text{ in-lb}$$

Moments could also be taken about *B* using the moments of the components of −**F**.

$$C = -(F_y \times 10)$$
$$= -(80 \sin 25° \times 10)$$
$$= -338 \text{ in-lb}$$

Although the problem cannot be solved completely with graphics, either the distance *d* or the component F'_y can be checked graphically, as shown in Fig. 3-23.

PROBLEMS

3-28. Two forces of 75 lb each are parallel, opposite in direction, and 8 in. apart. What is the magnitude of the couple they form?

3-29. A die and holder used for making threads on pipe and rods are shown in Fig. P3-29. If 15 lb is exerted on each handle as shown, what is the couple applied to the die?

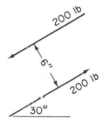

Figure P3-30

must transmit a torque of 3000 ft-lb, determine the force which must be carried by each of the four bolts in the coupling as indicated in Fig. P3-31(b).

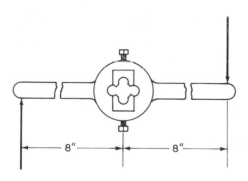

Figure P3-29

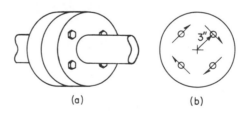

(a)   (b)

Figure P3-31

3-30. Find the magnitude of the couple shown in Fig. P3-30.

3-31. A coupling for a drive shaft is shown in Fig. P3-31(a). If the shaft and coupling

3-32. Find the sum of the two couples shown in Fig. P3-32.

3-33. Find the resultant couple for the forces shown in Fig. P3-33.

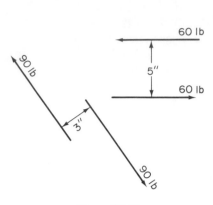

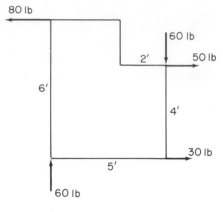

Figure P3-32   Figure P3-33

3-10. RESULTANTS OF NONCONCURRENT FORCE SYSTEMS

Only a few force systems are plane, concurrent force systems. Although we are omitting three-dimensional systems in this chapter, there are still two types of plane, nonconcurrent force systems to be considered. These are illustrated in Fig. 3-24, where Fig. 3-24(a) shows a plane parallel force system, and Fig. 3-24(b) shows a plane, nonconcurrent, nonparallel force system. Before determining the procedure for finding the resultant of such systems, we should first take another look at the definition of resultant of a system of forces. It is the simplest system of forces which can replace the given system of forces. The resultant may be one of the following simple systems: (1) a single force, (2) a single couple, or (3) a single force and a single couple. The resultant must cause the body on which it acts to behave in the same manner as it would behave under the original force system. In other words, if the original system tended to cause the body to move in a straight line, then the resultant must also tend to cause the same kind of motion. If the original system of forces tended to cause the body to rotate about some point, then the resultant must have the same tendency to cause rotation. The resultant could also tend to cause some combination of rotation and straight-line motion if this is the type of motion that the initial force system tended to cause.

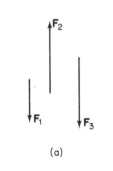

(a)

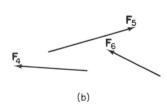

(b)

Figure 3-24

3-11. RESULTANT OF PLANE PARALLEL FORCES

Finding the magnitude of the resultant for such a system is very simple. All we have to do is add the forces algebraically, since they are parallel. However, the resultant must also cause the same tendency to rotate as the initial force system. To meet this requirement, the resultant force must have the same moment about any point as the initial system of forces had about the same point. Consequently, it is necessary to locate the resultant so that the above condition is met.

Mathematically, to find the resultant of a system of forces such as shown in Fig. 3-25, we must solve the following two expressions:

$$R = \sum F_i = F_1 + F_2 + F_3 + \cdots \qquad (3\text{-}6)$$

$$Rx = \sum (F_i d_i) = F_1 d_1 + F_2 d_2 + F_3 d_3 + \cdots \qquad (3\text{-}7)$$

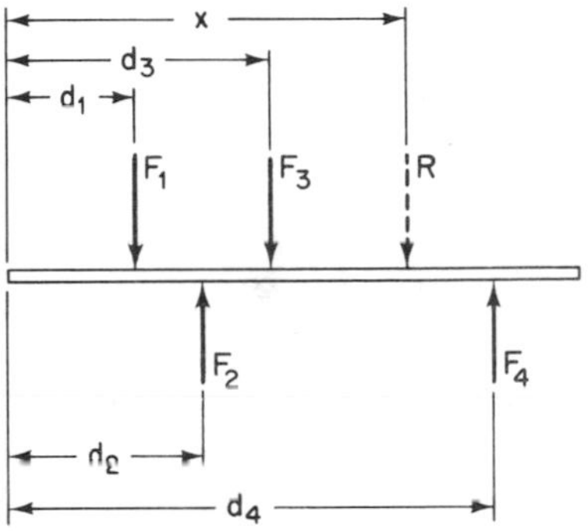

The last expression is solved for x, which is the distance from some convenient reference point to the location of the resultant. The distances, d_1, d_2, d_3, etc., are measured from the same reference point to the respective forces. When solving these expressions, you must remember our sign conventions for forces and moments. Positive forces point vertically up or horizontally to the right. Positive moments are counterclockwise. The terms in our moment equation depend on the direction of the moment and not on the direction of the force. The solution of a typical problem is illustrated in Ex. 3-7.

Figure 3-25

3-12. RESULTANT OF NONPARALLEL, NONCONCURRENT FORCES

Three forces are shown acting in Fig. 3-26(a) which are neither parallel nor concurrent. Forces $\mathbf{F}_2$ and $\mathbf{F}_3$ will intersect at point P. However, $\mathbf{F}_1$ will not intersect at P but at some other points with each of the other forces.

The force part of the resultant of the three forces can be obtained quite easily using the vector polygon for finding resultants. The location of the resultant must also be obtained. Again the moment of the resultant about any point must be the same as the sum of the moments of all the forces about

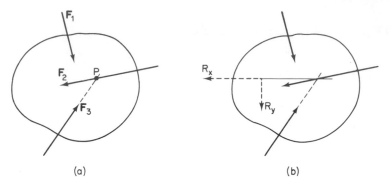

Figure 3-26   (a)   (b)

that same point. Mathematically we have

$$\mathbf{R} = \sum \mathbf{F}_i = \mathbf{F}_1 + \mathbf{F}_2 + \mathbf{F}_3 + \dots \qquad (3\text{-}8)$$

and

$$Rd = \sum (F_i d_i) = F_1 d_1 + F_2 d_2 + F_3 d_3 + \dots \qquad (3\text{-}9)$$

where

R is the resultant of all the forces,

d is the perpendicular distance from a reference point to the line of action of the resultant, and

d_1, d_2, d_3, etc. represent the corresponding perpendicular distances from the same reference point to the line of action of their respective forces.

By careful choice of reference points, it is possible to keep the solutions fairly simple. A convenient reference point is sometimes the point of concurrency of two of the forces, such as point P in Fig. 3-26(a). If the resultant is divided into two components, it can usually be positioned so that either the vertical or the horizontal component passes through the reference point. Then it is necessary only to find the distance to the other component. These ideas are illustrated in Fig. 3-26(b) as well as in an example problem to follow.

Resultants can also be obtained using a graphical method. The force vectors must be drawn to scale on a space diagram in their proper position on the body. The distances between forces must also be to scale. Fig. 3-27(a) shows such a figure with four forces on it. To solve the problem, it is helpful to use *Bow's notation* to label the vectors. As shown in the space diagram, Fig. 3-27(b), the spaces between the lines of action of the vectors are labeled, in sequence, using lower case letters. The vector polygon is then drawn as shown in

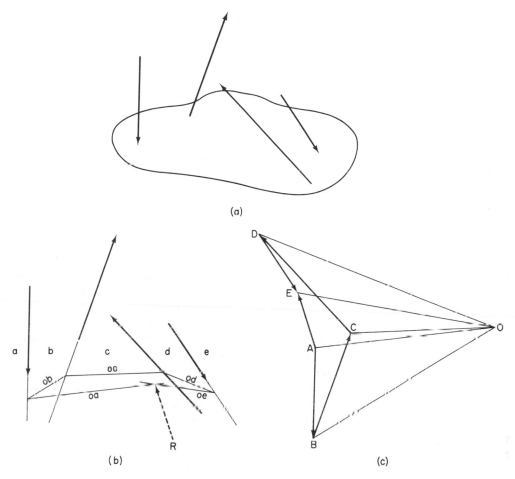

(a)

(b)   (c)

Fig. 3-27(c) with the vector *AB* being the vector bounded by   **Figure 3-27**
spaces *a* and *b*. It is placed at some convenient location to the
side of the space diagram. The rest of the vectors are added
consecutively. Note that the vectors must be drawn in alpha-
betical sequence for this type of graphical solution, even though
in general the order of addition is not important. The vectors
are labeled in upper case letters according to the labels of the
spaces on each side of the vectors in the space diagram. The
resultant is the vector from the origin, point *A*, to the tip of
the last vector, point *E*, in the vector polygon.

A pole point is then chosen at some point *O* in the
diagram usually located so that the rays, the lines drawn to
join *O* to the ends of the vectors, do not overlap. From some
convenient point on the line of action of *AB* in the space dia-

gram a line *ob*, parallel to *OB*, is drawn to intersect vector *BC*. From this point of intersection, a line *oc* is drawn to intersect the projection of vector *CD*. This procedure is continued until lines have been drawn in the space diagram parallel to all the rays. The polygon formed is called the string polygon. The lines parallel to the rays *OE* at the tip and *OA* at the tail of the resultant will intersect. This point of intersection is on the line of action of the resultant. Its direction and magnitude are obtained from the vector polygon.

The resultant for a plane, parallel force system may be found using the above method, since such a system is actually just a special case of the general plane, nonconcurrent force system as described above.

There are two special situations which may occur in solving problems graphically. If the vector polygon closes without a resultant and the string polygon in the space diagram does not close, then the resultant is a couple. In this case the first and last rays coincide in the vector polygon and will be parallel in the string polygon. The magnitude of the couple will be the product of the last ray in the vector polygon and the perpendicular distance apart of the parallel lines in the string polygon. Of course, both the ray and the distance must be to scale.

The second special case occurs when the string polygon closes and the vector polygon also closes. In this case the resultant of the force system is zero.

EXAMPLE 3-7

Find and locate the resultant of the three forces shown in Fig. 3-28.

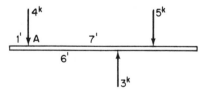

Figure 3-28

Since the forces are parallel, the resultant is the algebraic sum of the forces.

$$R = \sum F = -4 + 3 - 5$$
$$= -6 \text{ kips}$$

The moment of all the forces about some point must equal the moment of the resul-

$$\sum M_A:$$
$$Rx = \sum (F_i d_i)$$

tant about the same point. Moments are taken about A to eliminate the moment of the 4-kip force.

$$-6x = 3 \times 5 - (5 \times 7)$$

$$x = \frac{15 - 35}{-6} = 3.33 \text{ ft}$$

Resultant is 3.33 ft right of A

Graphically, the space or load diagram is drawn to scale as shown in Fig. 3-29(a). The vector polygon is a straight line, and because of the overlap of vectors, care must be taken to label the vectors clearly. The rays have been drawn in sequence, with the intersection of *oa* and *od* used to define the location of the resultant.

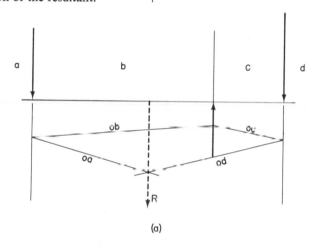

(a)

$R = 6^k$ at 3.33 ft. right of A

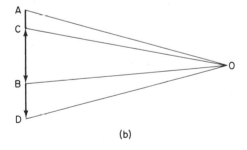

(b)

Figure 3-29

EXAMPLE 3-8

Find and locate the resultant for the three forces shown in Fig. 3-30(a).

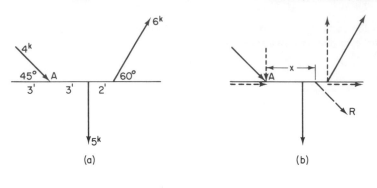

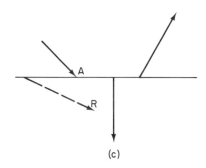

(c)

Figure 3-30

The components of the forces are shown in dashed lines in Fig. 3-30(b). The resultant can be obtained by summing the x and y components of the forces and finding the resultant of the two component sums by the Pythagorean theorem or the slide rule method. The assumed resultant is shown in Fig. 3-30(b).

$$\Sigma F_x = 4 \cos 45° + 6 \cos 60°$$
$$= 2.83 + 3.00$$
$$= 5.83 \text{ kips}$$
$$\Sigma F_y = -4 \sin 45° - 5 + 6 \sin 60°$$
$$= -2.83 - 5.00 + 5.20$$
$$= -2.63 \text{ kips}$$
$$R = 6.39 \text{ kips} \qquad 24.3°$$

To find the value of x where the resultant passes through the x axis, moments are taken about point A. The moment of the resultant about A must be the same as the sum of the moments of all the applied forces about point A. Notice that, as drawn in Fig. 3-30(b), the x components of the forces and the resultant will not have a moment about point A. The negative sign in the answer means that the resultant passes through the x axis to the left of the reference point A. The actual resultant is shown as a dashed line in Fig. 3-30(c).

$$\Sigma M_A:$$
$$R_y x = \Sigma (F_i d_i)$$
$$-2.63x = -(5 \times 3) + 5 \times 6 \sin 60°$$
$$= -15 + 26.0$$
$$= 11.0$$
$$x = \frac{-11.0}{2.63} = -4.18 \text{ ft}$$

The resultant passes through the x axis 4.18 ft left of A.

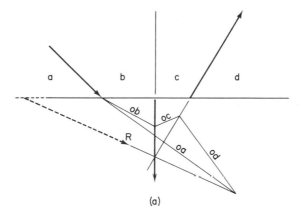

(a)

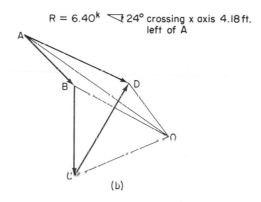

$R = 6.40^k \ \sqsubset 24°$ crossing x axis 4.18 ft. left of A

(b)

Figure 3-31

In the graphical solution, project the lines of action of the forces before labeling the spaces as shown in Fig. 3-31(a). Having all the known forces on the same side of the axis will help give the resultant force its correct orientation. The vector polygon is shown in Fig. 3-31(b), and the resultant is the vector *AD* required to close the polygon. In selecting the pole point *O* to form the rays, be careful that the angles formed are not so small that it becomes difficult to determine the points of intersection when the rays are transferred back to the string polygon. The starting vector in the vector polygon is *AB*. The starting string in the string polygon is *ob*, and it is started from the point where force *AB* intersects the *x* axis. The resultant **R** whose line of action has been obtained from the vector polygon, must pass through the intersection of strings *od* and *oa*.

PROBLEMS

3-34. Determine the magnitude and location of the resultant of two forces **A** and **B**, if **A** is 500 lb down, and **B** is 300 lb down and 4 ft to the right of **A**.

3-35. Determine the magnitude and location of the resultant of two forces **A** and **B**, if **A** is 5 kips down, and **B** is 2 kips up and 6 ft to the right of **A**.

3-36. Find and locate the resultant of the three forces shown in Fig. P3-36.

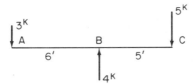

Figure P3-36

3-37. Find the magnitude and location of the resultant of the three forces shown in Fig. P3-37.

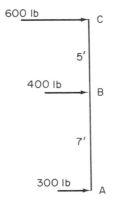

Figure P3-37

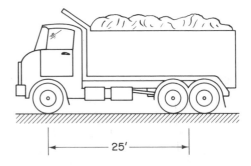

Figure P3-38

3-38. A large truck as shown in Fig. P3-38 has too long a wheelbase to fit on a weigh scale, so it is weighed by weighing the loads from the front and rear wheels separately. If the front wheels exert an 8-ton force and the rear wheels exert a 34-ton force on the scale, determine the total weight of the truck and the distance from the front wheels to where the total weight acts.

3-39. In designing a freight cart for use in a factory, the wheelbase *l* should be made fairly short so that it is easy to turn corners. However, if there is too much overhang, the cart will tip if a load is placed on the front. The resultant of the weight of the cart (300 lb) and the assumed load (200 lb) at the front must be at the front wheel, or to its right. Determine the minimum value of *l* for which the resultant of the two forces would be located at the front wheel. The cart is shown in Fig. P3-39.

3-40. Find and locate the resultant of the force system shown in Fig. P3-40.

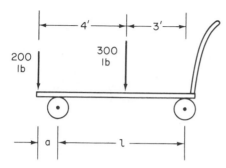

Figure P3-39

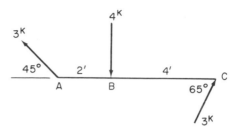

Figure P3-42

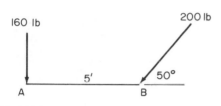

Figure P3-40

3-41. Find and locate the resultant of the force system shown in Fig. P3-41.

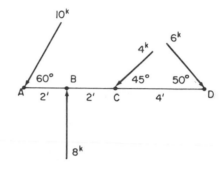

Figure P3-43

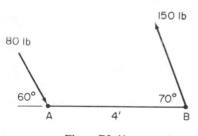

Figure P3-41

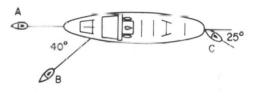

Figure P3-44

3-42. Find and locate the resultant of the force system shown in Fig. P3-42.

3-43. Find and locate the resultant of the force system shown in Fig. P3-43.

3-44. A steamship is maneuvered in a harbor by means of tugs as shown in Fig. P3-44. Tug *A* pulls with 20,000 lb, tug *B* pulls with 17,000 lb, and tug *C* pushes with

12,000 lb. The ship is 360 ft long. The rope from tug *B* is tied to a point 80 ft from the bow on the ship's centerline. Determine the resultant force exerted by the three tugs, and where the resultant force intersects the ship's centerline.

3-45. Part of a front-end loader in the process of raising the load is shown in Fig. P3-45(a), and the forces acting are shown in Fig. P3-45(b). Find the resultant of the forces shown.

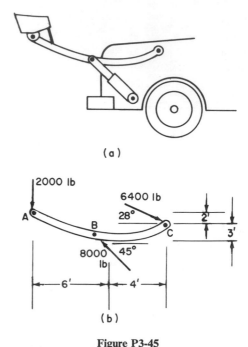

(a)

2000 lb

6400 lb

28°

A

B

C

2'

3'

8000 lb   45°

6'   4'

(b)

Figure P3-45

3-13. REPLACING A FORCE BY A FORCE AND A COUPLE

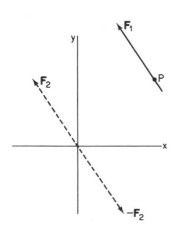

Figure 3-32

There are times when the above methods of finding resultants do not produce the most convenient form of resultant. This is true when it is desirable to have the resultant applied at some particular point. This particular point will seldom be the one that the resultant should pass through in order to produce the required moment.

Let us consider the simplest case of the single force F_1 passing through some point, P, as shown in Fig. 3-32. As we will discover in Chapter 9, it may make certain problem solutions simpler if the force is applied at some other point, such as the origin. The new force at the origin must have the same effect as the initial force. The following procedure achieves this result.

At the origin, place a second force F_2 which is identical to F_1 except for point of application. This, of course, changes the force system, which is an undesirable result. To overcome this, add $-F_2$ at the origin. Now $F_2 - F_2$ gives zero, so that we now have the same force system we started with. However, we do not discard $\mathbf{F}_2$ and $-\mathbf{F}_2$. Instead, look at $-\mathbf{F}_2$ and $\mathbf{F}_1$. They are parallel, equal in magnitude, and opposite in direction, and hence they constitute a couple. Thus we now have a force at the origin and a couple composed of $-\mathbf{F}_2$ and $\mathbf{F}_1$.

This procedure can be used with systems of several forces, simply by replacing each force with its equivalent at the desired location and its corresponding couple, and then obtaining the vector sum of all the forces and couples.

EXAMPLE 3-9

Replace the two forces shown in Fig. 3-33(a) with a single force at the origin and a couple.

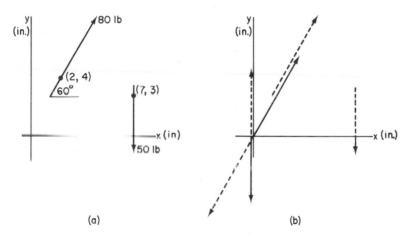

(a)   (b)

Figure 3-33

The two forces are shown at the origin in Fig. 3-33(b). The resultant is obtained by summing components and using the slide rule to find the magnitude and direction.

$$\Sigma F_x = 80 \cos 60°$$
$$= 40 \text{ ib}$$

$$\Sigma F_y = 80 \sin 60° - 50$$
$$= 69.3 - 50$$
$$= 19.3 \text{ lb}$$

$$R = 44.4 \text{ lb} \qquad 25.8°$$

Referring to Fig. 3-33(b), moments of the forces are taken about the origin. Using components saves calculating the perpendicular distance from the 80-lb force to the origin.

$$C = -(4 \times 80 \cos 60°) + 2 \times 80 \sin 60°$$
$$- (7 \times 50)$$
$$= -160 + 138.6 - 350$$
$$= -371 \text{ in-lb}$$

Although there is not a graphical method for obtaining couples, the resultant force can be checked using the vector polygon, as shown in Fig. 3-34.

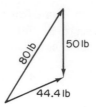

Figure 3-34

PROBLEMS

3-46. A 50-lb force is directed horizontally to the left at 3 in. above the x axis. Replace this force by a force at the origin and a couple.

3-47. An 80-lb force is directed vertically up and passes through point B with coordinates $(5, 0)$. Replace the force by a single force at A and a couple. The coordinates of A are $(2, 4)$. Coordinates are in feet.

3-48. Figure P3-48 shows a side view of a column with a bracket for supporting a beam. Replace the 12-kip force which the

beam exerts and the 8-kip force at the center of the column by a single force at the center of the column and a couple.

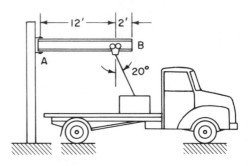

Figure P3-49

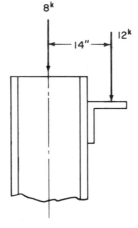

Figure P3-48

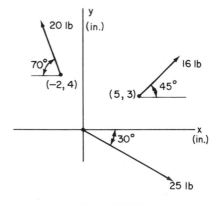

Figure P3-50

3-49. A small hoist at a loading dock is shown in Fig. P3-49. The weight of the boom *AB* is 1500 lb and acts down at the center of the boom, and the tension in the cable is 3000 lb. To design the connection at *A* it is necessary to replace the two forces acting on the boom by a single force acting at *A* and a couple. Determine the force and couple required.

3-50. Replace the three forces shown in Fig. P3-50 by a single force at the origin and a couple.

3-51. The system shown in Fig. P3-51 consists of two forces and a couple. Replace this system with a single force at the origin and one couple.

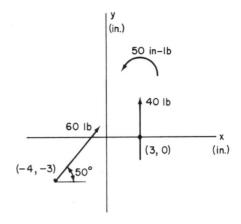

Figure P3-51

4

EQUILIBRIUM

Once you have mastered the use of vectors and can find components, resultants, moments, and couples, you are prepared to start solving real technical problems. Most of the structures we deal with are in an equilibrium condition. Determination of the forces required to maintain this equilibrium condition is an important part of any type of analysis problem associated with any form of mechanical or structural design.

4-1. BASIC PRINCIPLES OF EQUILIBRIUM

A body is said to be in equilibrium if both of the following conditions are met: (1) the resultant of all the externally applied forces on the body is zero, and (2) the sum of the moments about every point in the body is zero. In general, bodies in equilibrium are either at rest or moving at constant speed in a straight line.

Mathematically, a body in equilibrium must satisfy both of the following two expressions:

$$\sum \mathbf{F} = 0 \qquad (4\text{-}1)$$

$$\sum \mathbf{M} = 0 \qquad (4\text{-}2)$$

These equations are frequently expressed in terms of their rectangular components:

$$\Sigma F_x = 0$$
$$\Sigma F_y = 0$$
$$\Sigma F_z = 0$$

Fig. 4-1. The vertical lift bridge shown represents the efforts of many designers applying the principles of mechanics to the solution of a complex equilibrium problem. (Photo courtesy of Michigan Department of State Highways.)

$$\sum M_x = 0$$
$$\sum M_y = 0$$
$$\sum M_z = 0$$

Equilibrium problems usually involve situations where we know some of the forces on a body and need to determine the magnitude and/or direction of some unknown force or forces. As indicated above in the case of a three-dimensional problem, there are six equilibrium equations which will then permit us to solve for as many as six unknowns. For two-dimensional problems, there are only three equations which can be used: $\sum F_x = 0$, $\sum F_y = 0$, and $\sum M_z = 0$. This limits us to finding a maximum of three unknowns for two-dimensional or plane problems. (It should be noted that although we are limited to three equations for the plane problems, they will not necessarily be in exactly the form indicated.)

For any case in which we find there are more unknowns than we have equations of equilibrium available to solve for the unknowns, we have a *statically indeterminate* problem. This simply means that our equilibrium problems cannot be solved using statics only. In addition we must use relations depending on the deformation of the bodies. Thus, the statically indeterminate problems will be studied in a later course.

The lift bridge shown in Fig. 4-1 is one of countless design problems where the design depends on an understanding of the principles of equilibrium. By the time you have completed the study of this chapter, you should understand some of the analysis of forces in a structure as complex as the bridge shown. The analysis or determination of the forces in the members is one of the first steps towards design.

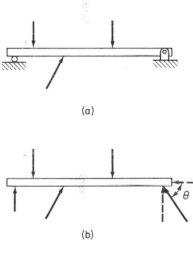

(a)

(b)

Figure 4-2

4-2. FREE-BODY DIAGRAMS

To solve equilibrium problems, we must learn to use an exceedingly helpful tool for problem solution. This tool is

                                                Equilibrium

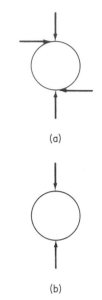

(a)

(b)

Figure 4-3

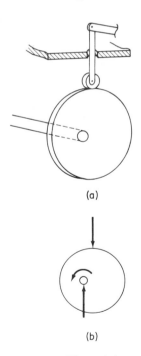

(a)

(b)

Figure 4-4

called a free-body diagram. A free-body diagram shows the body we are investigating with its supports removed and replaced by the forces which the supports exerted *on* the body, as well as showing all the other external forces acting *on* the body. The word *on* cannot be overemphasized. It is crucial that all forces acting on the body be shown, but no force acting in the body, or forces which the body exerts, are to be shown. A beam with three forces acting on it is shown in Fig. 4-2(a).

To make a free-body diagram of the beam, we would draw the beam without supports but replace the supports with the forces they exert on the beam. At the left end there is a roller which is assumed to be frictionless and which pushes up on the beam. Because the roller is frictionless, the only force it exerts is normal to the surface at the point of contact. We assume that the roller keeps the left end from falling down, so that the force exerted by the roller must then push up. If the roller exerted any horizontal force on the beam the beam would have to push back on the roller with a horizontal force. A free-body diagram of a roller under such circumstances is shown in Fig. 4-3(a). You can see that, if there is a horizontal force on the roller, the roller would start to roll, and thus it would not be in equilibrium. The free-body diagram of the roller as it actually exists for the given beam is shown in Fig. 4-3(b).

The right end of the beam is supported by a pin. A pin connection such as that shown can prevent motion in any direction in the plane of the paper. Although we do not know just what direction the force from the pin will have, we do expect that a force must be there, and that it must hold up the right end of the beam. It must also prevent the beam from moving to the right under the influence of the force with the horizontal component. Sometimes such unknown forces are shown as unknown rectangular components, as illustrated by the dashed lines in Fig. 4-2(b).

A cam and follower is shown in Fig. 4-4(a). A free-body diagram of the cam is shown in Fig. 4-4(b). The shaft applies a couple (frequently called a torque when transmitted by a shaft) to the cam. The follower exerts a force normal to the point of contact (assuming that the roller's action on the follower is frictionless), and the shaft must also exert an upward force to oppose the force exerted by the follower.

4-3. CONVENTIONS AND ASSUMPTIONS USED IN FREE-BODY DIAGRAMS

In drawing free-body diagrams, there are certain assumptions and conventions, or agreements, that are generally made. We usually neglect the weight of the member or parts, as we have done in the free-body diagrams discussed above. Actually, we are assuming that these parts are weightless, which is obviously not true. However, if their weight is small relative to the loads applied, our assumption of weightlessness simplifies our calculations and introduces only very small

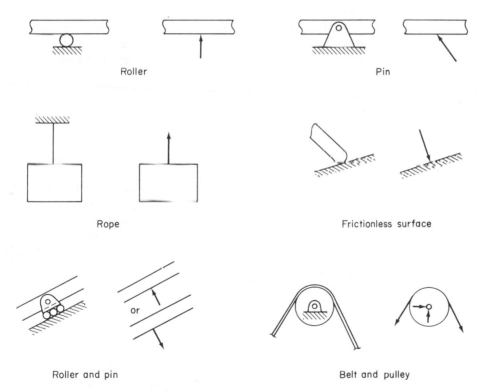

Roller   Pin

Rope   Frictionless surface

Roller and pin   Belt and pulley

errors. If you are to include the weight in your problem calcu-   **Figure 4-5**
lations, there will be some indication of this in the problem statement. In practice, whether or not the weight of the body or part is included in the calculations is a matter of judgment. We also assume that most surfaces in contact are frictionless.

This is not true either, but again, it generally simplifies calculations without introducing a very large error. If friction needs to be considered, this will usually be indicated clearly.

While on the subject of assumptions, we might as well realize that most of the loads shown as applied to a structure are simply the results of assumptions or educated guesses. In design work, the size of these assumed loads is based on considerable experience or is determined by design codes—laws which regulate many aspects of design. Nonetheless, whether the load assumed is based on considerable experience or law, it is still an assumption. Thus, if we make further simplifying assumptions, such as neglecting weight or friction, the degree of inaccuracy will not necessarily be increased significantly.

The conventions to which we have referred are illustrated in Fig. 4-5. Various types of supports are shown, as are the assumed reactions of these supports on the free body.

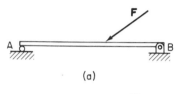

(a)

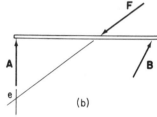

(b)

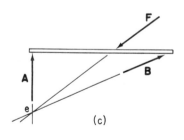

(c)

(d)

Figure 4-6

4-4. CONCURRENT COPLANAR FORCE SYSTEMS

The simplest type of equilibrium problem involves a concurrent, coplanar force system. This is a force system in which the lines of action of all forces are in one plane and pass through a common point. In such a system it is usually not necessary to consider moments, since any unknown force must also pass through the point of concurrency. Thus the equilibrium equation which we will use most often is $\sum \mathbf{F} = 0$. For convenience this will most frequently be written in the component form, $\sum F_x = 0$ and $\sum F_y = 0$. This gives us two equations to work with and thus limits us to finding a maximum of two unknowns. The unknowns will likely be one of the following combinations: an unknown force and direction, two unknown forces, or two unknown directions.

It may not be very apparent, but every body in equilibrium with only three nonparallel forces acting on it is a body which is being acted on by a concurrent force system. A beam with a single force $\mathbf{F}$ on it is shown in Fig. 4-6(a). When we draw the free-body diagram of the beam, we see that there are three forces as indicated in Fig. 4-6(b). The direction of the force at A must be vertical because of the roller, and the direction of the force on the pin at B is unknown. Note that the lines of action of the force $\mathbf{F}$ and the reaction $\mathbf{A}$ must intersect at point *e*. If moments are taken about point *e*, the force **B** will have a moment about point *e*, unless the line of action of the force **B** also passes through *e*. Thus the only way the body can be in equilibrium is if all the lines of action of the three forces pass through a common point of intersection. In the illustration the point of intersection must be *e*, since the lines of action of **F** and **A** are fixed and can intersect at *e* only. This permits us to find the line of action of all forces in this problem by means of trigonometry. If the lines of action are known, then the rest of the solution may be obtained by setting the sum of the forces equal to zero.

Equilibrium problems where the forces are concurrent can be solved very easily using a graphical method. Generally, the known forces are drawn to form a vector triangle or polygon, with the forces tip to tail. If the body is in equilibrium, the polygon or triangle must close. Any additional force required to close the triangle or polygon is the force required to maintain equilibrium in the body. This is shown in Fig. 4-6(d). The magnitude and direction of the force can be obtained by scaling from the figure. Of course, if the unknown quantities are the magnitude of one force and the direction of another force, the problem then becomes slightly more complex and calls on a bit of your drafting ingenuity.

EXAMPLE 4-1

A 200-lb block is supported by two cables as shown in Fig. 4-7(a). Calculate the force in each of the two cables.

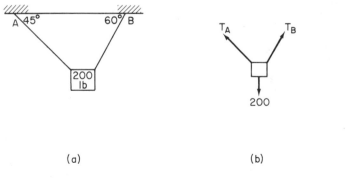

Figure 4-7

If we are to find the tensions in cables *A* and *B* using the principles of equilibrium, we must draw a free-body diagram in which we replace the cables by the forces in them as shown in Fig. 4-7(b). We note that we have two unknown forces and that we know the directions of the two forces. To find the unknown forces, first we mentally check to see if either of the equilibrium equations will be an equation with one unknown. Since both equations do have two unknowns, we must set up the two equilibrium equations and solve them simultaneously.

$\sum F_y = 0$

$T_A \sin 45° + T_B \sin 60° - 200 = 0$

$.707T_A + .866T_B - 200 = 0$ (a)

$\sum F_x = 0$

$-T_A \cos 45° + T_B \cos 60° = 0$

$T_B = \frac{.707T_A}{.5} = 1.414T_A$ (b)

Substituting for T_B in Eq. (a),

$.707T_A + .866 \times (1.414T_A) - 200 = 0$

$.707T_A + 1.227T_A = 200$

$1.934T_A = 200$

$T_A = \frac{200}{1.934} = 103.4 \text{ lb}$

Substituting for T_A in Eq. (b),

$T_B = 1.414 \times 103.4 = 146.1 \text{ lb}$

The graphical solution is shown in Fig. 4-7(c). The length and direction of the known 200-lb force is drawn to scale, and lines parallel to the known direction of T_A and T_B are drawn from the tail and tip of the 200-lb force, respectively. Where these two lines intersect gives us the end points for the vectors T_A and T_B such that our vector equation $\sum F = 0$, or $T_A + T_B + W = 0$. The values of T_A and T_B are then scaled from our drawing. The values could also be checked by using the vector triangle and the sine law.

PROBLEMS

4-1. Two forces act on the body shown in Fig. P4-1. Find the third force required to keep the body in equilibrium.

4-2. Find the force necessary to maintain equilibrium in the body shown in Fig. P4-2.

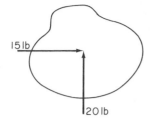

Figure P4-1

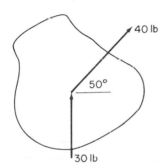

Figure P4-2

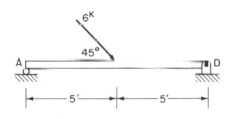

Figure P4-3

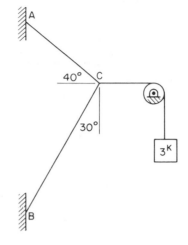

Figure P4-4

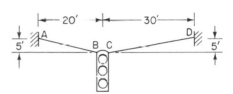

Figure P4-5

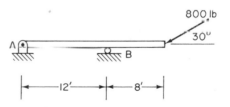

Figure P4-6

4-3. Determine the forces acting at the supports of the beam shown in Fig. P4-3.

4-4. Find the tension in ropes *AC* and *BC* of Fig. P4-4.

4-5. Figure P4-5 shows a traffic light suspended by wires. If the light weighs 50 lb, determine the tension in the two wires *AB* and *CD*.

4-6. Find the forces on the beam at the supports *A* and *B* for the cantilever beam shown in Fig. P4-6.

4-7. Determine the force **F** required for

equilibrium and the force in the cable *AB* shown in Fig. P4-7.

4-8. If the force in *BC* of Fig. P4-8 is 150 lb, find the magnitude and direction of the force in rope *AB*.

4-9. A beam is supported by two cables as shown in Fig. P4-9. Determine the tension in each cable and the angle θ. The beam is 32 ft long and its weight of 4 tons may be assumed to act at the center of the beam.

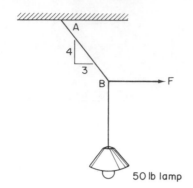

50 lb lamp

Figure P4-7

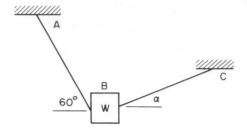

Figure P4-10

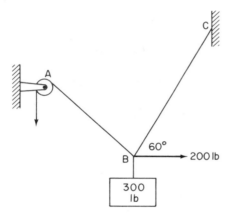

Figure P4-8

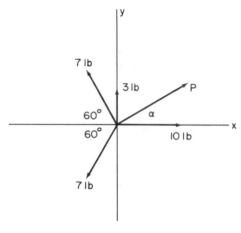

Figure P4-11

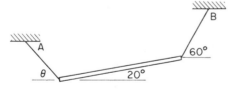

Figure P4-9

4-10. Figure P4-10 shows a weight *W* supported by two ropes. If the tension in *AB* is 200 lb and in *CB* is 400 lb, determine *W* and α.

4-11. Determine the magnitude and direc-

tion of the force **P** shown in Fig. P4-11 if the system is in equilibrium.

4-12. A ladder is shown leaning against a smooth wall in Fig. P4-12. Assume that the 18-1b weight of the ladder acts at the ladder's center. The lower end of the ladder rests on a rough surface, so that the direction of the force is unknown. Determine the reactions on the ladder at the top and bottom.

4-13. A strap is looped around a 10-in. square steel billet of weight *W* lb and attached to an overhead crane as shown in Fig. P4-13. Find the force in the strap.

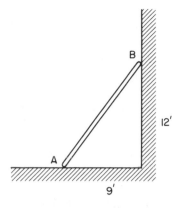

Figure P4-12

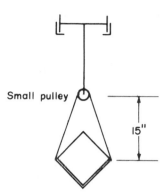

Figure P4-13

4-5. COPLANAR NONPARALLEL FORCE SYSTEMS

Force systems which are not concurrent require slightly more analysis than concurrent force systems but are certainly no more difficult. In this type of problem we can usually solve for up to three unknowns—unknown forces, directions, or some combination of forces and directions. To solve for these unknowns, we must use both of our equilibrium equations, $\Sigma \mathbf{F} = 0$ and $\Sigma \mathbf{M} = 0$. These will usually be used in their component form, so that most of the time we will be using the expressions $\Sigma F_x = 0$, $\Sigma F_y = 0$, and $\Sigma M = 0$. (It should be noted that it is possible to take moments about two different points and use only one of the equations summing forces. However, it is *not* possible to use four equations, i.e., two force equations and two moment equations. If you need this many equations to solve your problem, it is statically indeterminate, and by using the four equations you will probably just prove that $1 = 1$ or $0 = 0$.) Figure 4-8(a) is typical of a large number of the coplanar, nonparallel force systems which you might encounter. The beam has three known forces and two unknown reactions at A and B. On drawing the free-body diagram shown in Fig. 4-8(b), we see that not only are the forces at A and B unknown, but also the angle of the reaction at A is unknown. Much of the time it will be more convenient to treat this as a problem with three unknown forces A_x, A_y, and B_y. Forces A_x and A_y are shown in dashed lines. The angle θ can be found from A_x and A_y if it is needed.

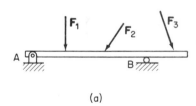

(a)

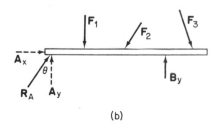

(b)

Figure 4-8

Once we have drawn our free-body diagrams, our problem is to determine the simplest method of finding the unknown reactions. This usually is to find equations we can use which will contain only one unknown term. Thus, before actually trying to solve the problems, we should first think through the possible equations to try to find one with only one unknown. Referring to Fig. 4-8(b), if we sum forces in the y direction, we will find that we have two unknowns in our equation, and if we sum forces in the x direction, we will have only one unknown in our equation. The third choice is to take moments about some convenient point. Although we can take moments about any point, if moments are taken about either A or B, we will have only one unknown in the equation. If we take moments about any other point on the beam, there will be two unknowns in our moment equation. The complete solution will include the magnitudes of all forces and their directions.

You will note that in setting up the free-body diagram it is necessary to assume the direction of some of the forces, that is, whether the x or y components are headed in the positive or negative directions. There is no need to spend a great deal of time in making your assumption. If you assume the wrong direction for a component, you will get a negative answer, which tells you that the correct direction is opposite to your assumed direction. It is sometimes suggested that, when assuming directions for components, you should always assume that they are positive. If this assumption is made, you will always get the correct sign associated with your answer, since a negative sign indicates a force in the negative direction and a positive sign indicates a force in the positive direction.

Some problems may be similar to that shown in Fig. 4-9, in which a couple acts in addition to the forces. Since the couple has a tendency to cause rotation, the reactions must be able to resist not only the moments of the forces but also the moments resulting from the couples. This type of problem cannot be solved conveniently using the graphical method outlined below.

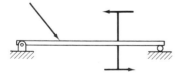

Figure 4-9

4-6. PARALLEL FORCE SYSTEMS

A parallel force system is just a special case of the co-planar force system. It will be simpler to handle because the forces are in one direction only. Consequently, one of our equilibrium equations will be of no value, and we will be limited to finding two unknowns.

4-7. GRAPHICAL PROCEDURE FOR COPLANAR FORCE SYSTEMS

The string polygon can be used to find reactions for nonconcurrent force systems as shown in Fig. 4-10. The

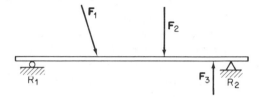

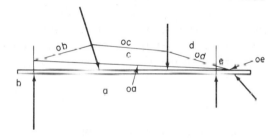

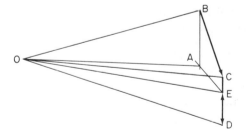

Figure 4-10

initial procedure is similar to that used for finding resultants; we draw a diagram of the beam to scale with the loads shown, using Bow's notation. A force polygon is then drawn complete with rays from some convenient pole. However, it cannot be completed, because the force required to close the diagram is actually the resultant of the two reactions. We leave the force polygon and return to making the string polygon. We should start constructing the string polygon from the point of intersection on the beam of the force whose magnitude and line of action are unknown: R_2 in Fig. 4-10(a). The lines in the string polygon are drawn parallel to the rays in the force polygon. The last ray (*ob* in the figure) will intersect the line of action of the other reaction. Note that this second reaction R_1 has a known direction. The two points of intersection, one on the line of action of each of the unknown forces, now specify the ends of the last line in the string polygon, so it can be drawn in to close the string polygon. There must be a ray in the force polygon parallel to this line. It will intersect the line of action of the unknown force R_1 which is AB in the force polygon and thus define the length of the force vector. The second unknown force R_2, which is AE in the force polygon, is then the force required to close the force polygon. With scale and protractor, it is now possible to determine the magnitude and direction of our two unknown forces.

This same approach may also be used for the special case of a system of parallel forces. Because the forces in the force polygon overlap to form a straight line, extra care is required in labeling the forces to prevent confusion.

EXAMPLE 4-2

Determine the reactions at A and B for the beam shown in Fig. 4-11(a).

For any equilibrium problem, the first step is to draw a free-body diagram as shown in Fig. 4-11(b). The reaction at B must be vertical as discussed previously. The pin at A could support a horizontal force component, but since all the loads are vertical, there is no need for a horizontal component of reaction at A.

To solve the problem we first mentally determine which of the equations of equilibrium will be the easiest to use. If we sum forces in the y direction, we will have two unknowns in our equation, A_y and B_y.

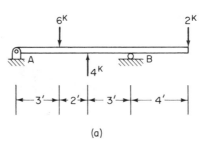

(a)

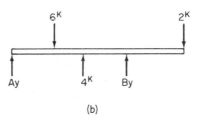

(b)

Figure 4-11

If we take moments about point A, the only unknown in our equation will be B_y.

$$\Sigma M_A = 0$$

$$-(3 \times 6) + (5 \times 4) + (8 \times B_y)$$
$$- (12 \times 2) = 0$$

$$-18 + 20 + 8B_y - 24 = 0$$

$$8B_y = 22$$

$$B_y = 2.75 \text{ kips}$$

Now if we sum forces in the y direction, there is only one unknown to solve for, A_y.

$$\Sigma F_y = 0$$

$$A_y - 6 + 4 + 2.75 - 2 = 0$$

$$A_y = 1.25 \text{ kips}$$

The graphical solution is shown in Fig. 4-12. The first step is to draw the beam and loads to scale as shown in Fig. 4-12(a). The spaces between the loads are labeled using Bow's notation. Note that in order to have all the known forces in sequence, they are labeled as if acting on the top of the beam, and the unknown forces are labeled as if acting on the bottom of the beam. The force polygon, in this case a straight line, is drawn, and a pole point O is chosen. Rays are drawn from the pole point to the ends of the vectors in the force polygon.

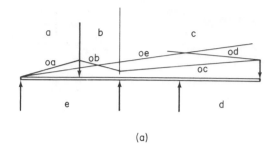

(a)

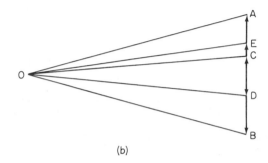

(b)

Figure 4-12

We then start drawing the string polygon, with the lines parallel to the rays. Start at a point on the line of action of one of the unknown forces where it meets the beam. In this case the string polygon was started at the left reaction with line *oa*. Lines *ob*, *oc*, and *od* may then be added to the string polygon, each meeting the previous line at the line of action of the force. Only one more line is required to close the polygon. It is *oe*, and both its ends are already determined, for one end must be where *oa* intersects the line of action of the left reaction, and the other end must be where *od* intersects the line of action of the right reaction. Now a ray *OE* is drawn in the force polygon parallel to *oe*. This determines the location of *E* in the force polygon and the two reactions can then be determined by scaling from the force polygon.

EXAMPLE 4-3

Find the reactions at the supports for the beam shown in Fig. 4-13(a).

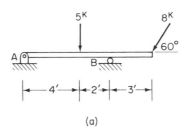

(a)

The first step in the analytical solution is to draw the free-body diagram as shown in Fig. 4-13(b). The support at B is frictionless and thus must be vertical. The support at A is a pin. This reaction is shown as two components: a vertical component, and a horizontal component to react against the horizontal component of the 8-kip force. Thus in the y direction, there are two unknown forces.

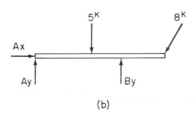

(b)

Figure 4-13

By taking moments about point A which is on the line of action of the left reaction, we can find B_y.

$$\Sigma M_A = 0$$
$$-(4 \times 5) + (6 \times B_y) - (9 \times 8 \sin 60°) = 0$$
$$6B_y = 20 + 62.4$$
$$B_y = \frac{82.4}{6} = 13.73 \text{ kips}$$

Now we can find A_y by summing forces in the y direction.

$$\Sigma F_y = 0$$
$$A_y - 5 + B_y - 8 \sin 60° = 0$$
$$A_y = 5 - 13.73 + 6.93$$
$$= -1.80 \text{ kips}$$

The negative sign means that we guessed wrong in assuming the direction of A_y in our free-body diagram.

$$A_y = 1.80 \text{ kips down}$$

We may now find A_x by summing forces in the x direction.

$$\Sigma F_x = 0$$
$$A_x - 8 \cos 60° = 0$$
$$A_x = 4.00 \text{ kips}$$

The resultant at *A* is found using the slide rule.

The total reaction at *A* is 4.38 kips   24.3°

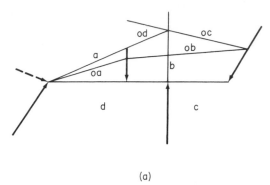

(a)

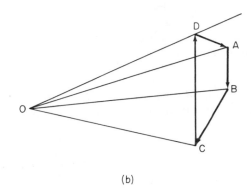

(b)

Figure 4-14

To solve the problem graphically, the beam and loads are drawn to scale as shown in Fig, 4-14(a). The spaces between forces are lettered using Bow's notation, keeping the two unknown forces in sequence. A force polygon is drawn as shown in Fig. 4-14(b), and rays are drawn from the pole point *O*. The strings are then drawn in the string polygon, starting at the point where the left reaction intersects the beam. The string polygon should start at a point on the line of action of the unknown about which we know the least, and for the left reaction we know neither the magnitude nor the direction. The strings *oa*, *ob*, and *oc* can be drawn without difficulty. The closing string must go from the intersection of *oc* on the line of action of the right reaction, to the starting point of the string polygon. The ray *OD* may then be drawn parallel to string *od*. Since force *CD* is vertical, the point of intersection of the ray *OD* and a vertical from *C* defines the length of *CD* and *DA*, whose lengths and directions can then be scaled.

PROBLEMS

4-14. A beam 12 ft long has a support at each end, and a load of 6 kips down at 4 ft from the left end and 8 kips down at 2 ft from the right end. Determine the reactions on the beam from the supports.

4-15. Two weights of 500 lb and 800 lb are placed on a 9-ft beam at distances of 2 ft and 6 ft, respectively, from the left end. If the beam is supported at each end, determine the reactions on the beam from the supports.

4-16. Determine the reactions on the beam at *A* and *B* for the beam shown in Fig. P4-16.

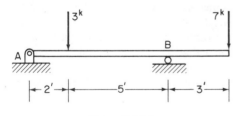

Figure P4-16

4-17. Find the force in cable *AB* which holds up the member *CD* in Fig. P4-17.

4-18. Find the reactions at *A* and *B* for the beam shown in Fig. P4-18.

4-19. Determine the reaction from the pin at *A* and from the roller at *B* in Fig. P4-19.

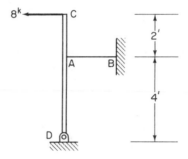

Figure P4-17

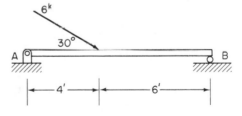

Figure P4-18

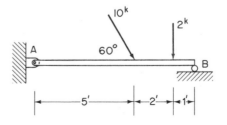

Figure P4-19

4-20. Determine the reactions at the two supports for the beam shown in Fig. P4-20.

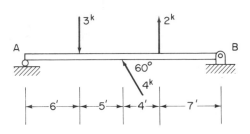

Figure P4-20

4-21. The member shown in Fig. P4-21 has a weight of 2000 lb which acts at the center of the member. Find the total reactions at A and B, if the wall at B is smooth.

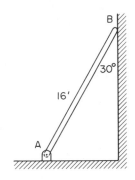

Figure P4-21

4-22. Find the reactions at A and B for the beam shown in Fig. P4-22.

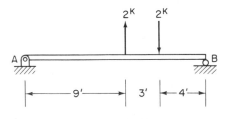

Figure P4-22

4-23. Determine the reactions on the beam shown in Fig. P4-23.

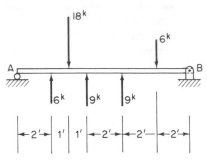

Figure P4-23

4-24. Determine the reactions at A and B for the bent bar shown in Fig. P4-24.

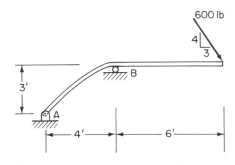

Figure P4-24

4-25. A 60-ft beam is being lifted by two cranes as shown in Fig. P4-25. The beam weighs 20 tons, and the weight acts through the center of the beam. Find the tension in the cables at A and B, and the angle θ. Note that since the beam is a member with three forces on it, you can also check the solution using the methods of Sec. 4-4.

4-26. A couple of magnitude 50 ft-lb is applied to the end of the bent beam as shown in Fig. P4-26. Find the reactions at A and B.

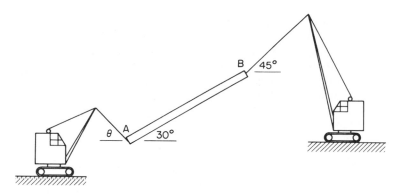

Figure P4-25

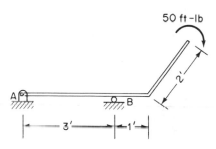

Figure P4-26

4-27. Determine the total reaction from the pin at *B* in Fig. P4-27.

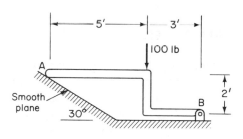

Figure P4-27

4-28. The bent bar shown in Fig. P4-28 is supported by a pin at *A* and a roller at *B*. Find the reactions at *A* and *B*.

4-29. The glass rod *AB* shown in Fig. P4-29 weighs 5 oz and is 6 in. long. It is placed in a glass tumbler *C* in a position of equilibrium similar to that shown. If the tumbler is 2.5 in. in diameter and if all surfaces are smooth, what is the angle θ?

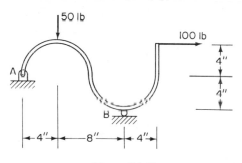

Figure P4-28

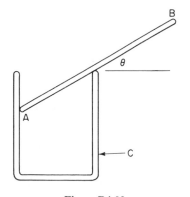

Figure P4-29

4-30. The pin at *A* of the structure for a traveling crane as shown in Fig. P4-30 is cracked and will not support more than 30 kips. The bent or truss on which the hoist travels weighs 25 kips. This weight acts through the center of gravity shown. The hoist with its load weighs 40 kips. What is the maximum distance from *A* that the hoist can travel without causing the pin at *A* to break?

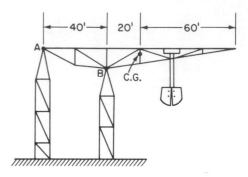

Figure P4-30

4-31. Determine the value for *x* if the beam shown in Fig. P4-31 is in equilibrium. The pulleys shown are frictionless and the beam is weightless.

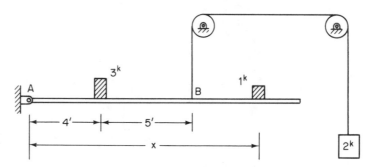

Figure P4-31

4-8. FRAMES AND MACHINES

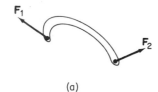

(a)

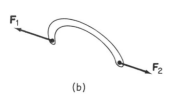

(b)

Figure 4-15

Analysis of the forces on the parts or component members of frames and machines represents an application of the use of the principles of equilibrium. A *frame* is a system of connected members designed to carry loads. A *machine* is a system of connected members designed to transmit forces. In addition to their different functions, it will be observed that a frame is usually rigid, whereas a machine is made up of members which can move relative to one another. The distinction is perhaps not very important for our purposes, since the method of analysis is essentially the same for both types.

Before trying to analyze any frames or machines, there is a special situation which we should consider. This is the *two-force member*. As might be expected from the term used, a two-force member is a member with two forces acting on it. Because this member must be in equilibrium, we can deduce a number of facts about the forces. Figure 4-15(a) shows a member with two forces acting on it. First, if we sum forces, the only way for $\Sigma\,\mathbf{F} = 0$ to be true is if $\mathbf{F}_1 + \mathbf{F}_2 = 0$. $\mathbf{F}_1$ must be equal in magnitude and opposite in direction to $\mathbf{F}_2$. The forces must also be colinear. If they are parallel, they will form a couple, and the member will not be in equilibrium, for $\Sigma\,M \neq 0$ if there is a couple acting on the member. Thus, for equilibrium, the forces on a two-force member must be (1) equal in magnitude, (2) opposite in direction, and (3) act along a line joining the two points of application of the forces, as shown in Fig. 4-15(b).

Developing an ability to recognize two-force members will assist in making analysis of many problems simpler.

4-9. ANALYSIS OF FRAME AND MACHINE PROBLEMS

Figure 4-16(a) shows a frame with two forces applied to it. To properly design any such frame or machine, it is necessary to determine all the loads applied to each member and connection so that the members and connections may be made sufficiently large to carry the loads safely, but not so large as to waste material. A logical first step in analyzing any such problem would be to draw a free-body diagram of the frame and try to determine the reactions at the supports. You will notice that in Fig. 4-16(b) there are four unknown reactions at the supports. The structure as a whole is statically indeterminate, but this does not mean that we are unable to solve the problem. If the frame is taken apart as shown in Fig. 4-16(c), and a free-body diagram is drawn of each part, then it will be possible to solve for all the forces exerted on each member of the frame.

Drawing these free-body diagrams requires some judgment and care. Attention must be paid to the type of reaction that can exist at each support. If there is anything such as friction, or the lack of friction, it will influence the direction or line of action of the force. Remember that if any of the members are two-force members, you will know the line of action of the force in them. Remember also that you must be consistent. If you assume that a force is headed in a particular direction, maintain your assumption on any other free body

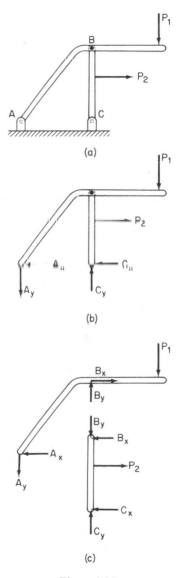

Figure 4-16

which shows the same force, and on the connecting bodies. Newton's third law, which states that for every action there is an equal and opposite reaction, must be carefully obeyed. This means that if member BC in Fig. 4-16(c) is pushing up at B on member AB, then in turn, member AB must be pushing down at B on member BC.

After drawing the free-body diagrams, it then becomes possible to solve for as many as three unknowns for each free body. Note that in Fig. 4-16 there is a total of three free-body diagrams, one of the entire frame, and one each of members AB and BC. The unknown forces are obtained by using the equilibrium equations $\sum F_x = 0$, $\sum F_y = 0$, and $\sum M = 0$.

EXAMPLE 4-4

Find the pin reactions at A and B for the frame shown in Fig. 4-17(a).

The first step is to draw a free-body diagram of the whole frame as shown in Fig. 4-17(b). Since neither AC nor DE are two-force members, we will not know the direction of the reactions at A and D, so we will show the x and y components of the forces. Since the force at B is internal to the frame, the frame must be taken apart so that the reaction at B becomes an external force as it appears in the free-body diagram of member AC. Note that if we consider the pulley free body, the tensions in both cables must be 800 lb. This can be proven by taking moments about E in the free body of the pulley. Thus the force at C on member AC will also be 800 lb. Both the free body of the frame and of member AC have four unknown forces acting on them. However, two of the forces, A_x and A_y, are common to both free-body diagrams.

If we consider the free body of member AC, we find that if we take moments about point A, we will get an equation with just one unknown.

In member AC

$$\sum M_A = 0$$

$$-(7 \times B_y) + 11 \times 800 = 0$$

$$B_y = \frac{11 \times 800}{7} = 1258 \text{ lb}$$

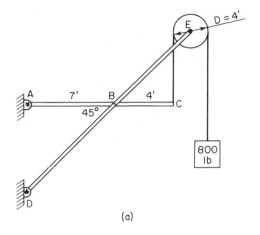

(a)

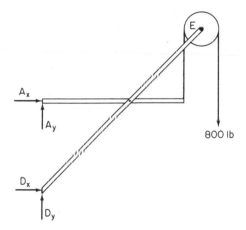

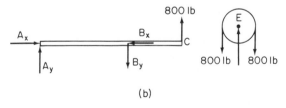

(b)

Figure 4-17

Now we have enough information to find A_y by summing forces in the y direction.

$$\sum F_y = 0$$

$$A_y - 1258 + 800 = 0$$

$$A_y = 458 \text{ lb}$$

We cannot find either A_x or B_x from the free-body diagram of AC. If we take moments about D in the free-body diagram of the whole frame, there will be only one unknown in the equation.

From the entire frame

$\Sigma M_D = 0$

$-(7 \times A_x) - (15 \times 800) = 0$

$A_x = -\dfrac{15 \times 800}{7} = -1715\ \text{lb}$

A_x is 1715 lb to the left

On the free-body diagram of member AC, there is only one force left which is unknown. The force B_x may be obtained by summing forces in the x direction.

In member AC

$\Sigma F_x = 0$

$A_x - B_x = 0$

$-1715 - B_x = 0$

$B_x = -1715\ \text{lb}$

B_x is 1715 lb to the right

The resultant reactions are found using the Pythagorean theorem or the slide rule method.

The reactions on member AC are

1770 lb   15° ⬎ at A and

2130 lb   36.2° ⬊ at B.

Notice that where negative signs were obtained for answers, the sign was kept in subsequent substitutions. Free-body diagrams must not be changed when an incorrect assumption is made, for then diagrams and equations would not agree.

Although this problem does not lend itself to a completely graphical solution, a check on the answer can be made graphically, as shown in Fig. 4-18. The member AC has been drawn to scale, and the forces at A, B, and C have been drawn in at their proper angles. As indicated, the three forces meet at a common point. Also, in Fig. 4-18(b) the three forces have been added vectorially, and their resultant has been found to be zero. Thus, since the forces as solved for are found to be concurrent and have a vector sum of zero, they may be correct. This procedure does not guarantee correctness, but it should help you to discover some glaring errors if they are made.

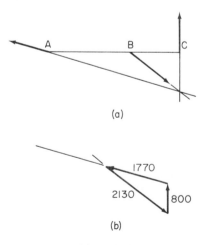

(a)

(b)

Figure 4-18

PROBLEMS

4-32. Determine the reactions at *A*, *B*, and *C* for the frame shown in Fig. P4-32.

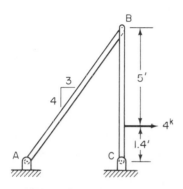

Figure P4-32

4-33. A rigid plane frame is shown in Fig. P4-33. Bars *AB* and *CB* are weightless, and the pins at *A*, *B*, and *C* are smooth. Compute the components of all pin reactions and the tension in the cord.

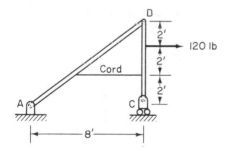

Figure P4-33

4-34. The rope holding the 2-kip load passes over a pulley at *E* and is attached to the frame at *B* as shown in Fig. P4-34. Determine the magnitude and direction of the total reaction at *D*, and the *x* and *y* components of the reaction at *A*.

4-35. A three-hinged arch is shown in Fig. P4-35. Determine the magnitude and direc-

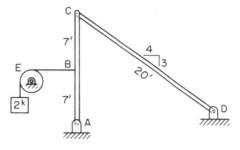

Figure P4-34

tion of the hinge reaction at *A* if the structure is considered to be weightless.

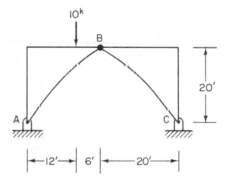

Figure P4-35

4-36. There are hinges at *A*, *B*, and *C* in the structure shown in Fig. P4-36. Determine the components of the reactions at *A* and *B*.

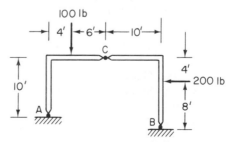

Figure P4-36

4-37. Figure P4-37 shows a clamping device with a hinge at *A* and a clamping screw at *C*. If the force on the piece being held at *B* is 40 lb, what is the tension in the screw at *C*?

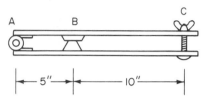

Figure P4-37

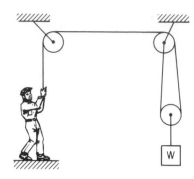

Figure P4-38

4-38. The weight shown in Fig. P4-38 is 160 lb, and the man weighs 200 lb. Determine the force that the floor exerts on the man.

4-39. Neglecting friction, determine the reactions of the two pipes shown in Fig. P4-39 at points *A*, *B*, and *C*. Each pipe weighs *W* lb.

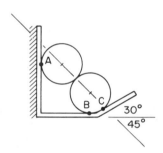

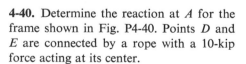

Figure P4-39

4-40. Determine the reaction at *A* for the frame shown in Fig. P4-40. Points *D* and *E* are connected by a rope with a 10-kip force acting at its center.

4-41. The beam *AC* shown in Fig. P4-41 weighs 400 lb, and its weight is assumed to act at its center. *BD* may be considered to be weightless. If the tension in BD is not to exceed 1000 lb, determine the maximum value for *W*.

4-42. Determine the magnitude and direction of the force on the pin of the block or pulley at *A* in Fig. P4-42.

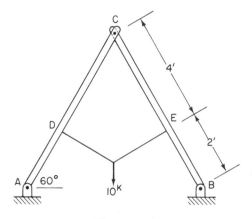

Figure P4-40

4-43. Find the total force P required on the end of the piston to maintain equilibrium for the machine shown in Fig. P4-43. Assume that there is no friction on the cylinder walls, and that the 700-lb load is in the middle of AB.

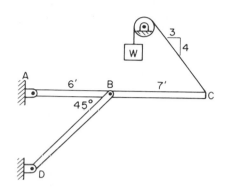

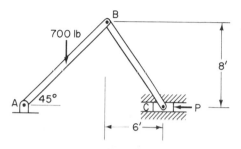

Figure P4-41

Figure P4-43

4-44. The trailer shown in Fig. P4-44 weighs 800 lb, and its cargo weighs 1200 lb. The composite center of gravity where their combined weight acts is shown. The combination of wind resistance and friction at the wheel R is 100 lb and is known to act through the center of gravity. If the trailer is in equilibrium, what is the force acting on the trailer at the ball socket?

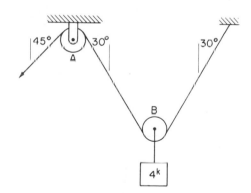

Figure P4-42

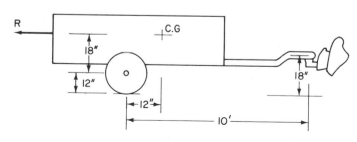

Figure P4-44

4-45. Find the value of the angle α when the force in the turnbuckle shown in Fig. P4-45 is $P/5$. Members AB, BC, CD, and AD are all the same length.

4-47. Figure P4-47 shows a 180-lb man standing on a 100-lb plank which is supported at one end by a pin and at the other end by a rope passing over a pulley. The weight of the plank acts at its center. Find the tension in the rope and the force the man exerts on the plank.

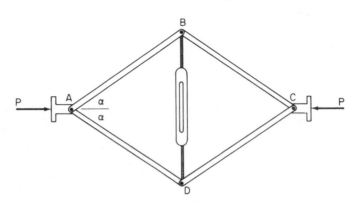

Figure P4-45

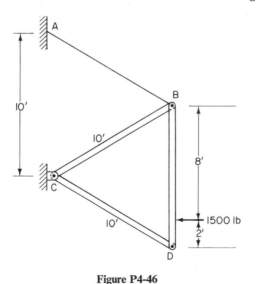

Figure P4-46

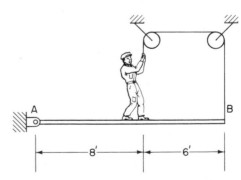

Figure P4-47

4-46. Find the force in the cable AB and the x and y components of the force at D in the frame shown in Fig. P4-46.

4-48. Determine the tension in cable AB and the reaction at C for the structure shown in Fig. P4-48. The pulleys have a small diameter.

4-49. For the frame shown in Fig. P4-49 determine the x and y components of the reaction at A.

4-50. The crane boom shown in Fig. P4-50 is pinned at A and supported by an hydraulic ram BC. If the boom AD has a weight of 3 kips acting at its center and the load supported is 5 kips, find the force in the ram. The diameter of the pulley at D may be neglected.

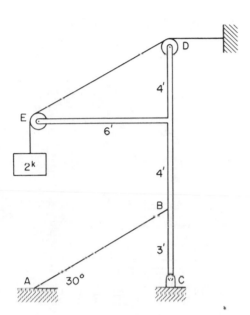

Figure P4-48

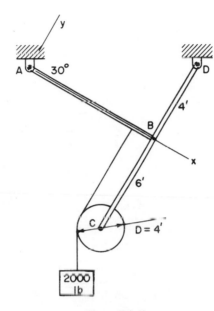

Figure P4-49

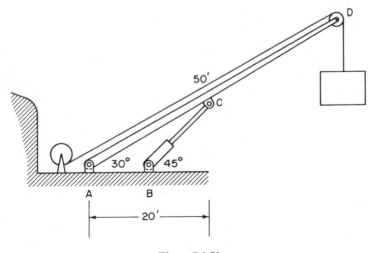

Figure P4-50

4-51. In Fig. P4-51, *A*, *B*, *C*, and *D* are pin connections. The force exerted on the block at *E* by the support is 80 lb. Determine the distance *x* so that the block remains in the position shown. *AB* and *CD* are weightless.

4-53. Members *AB* and *CD* are connected by a pin attached to *AB* which slides without friction in a slot in *CD*. Determine the value of the couple *M* applied to *AB* required to maintain equilibrium in the structure shown in Fig. P4-53.

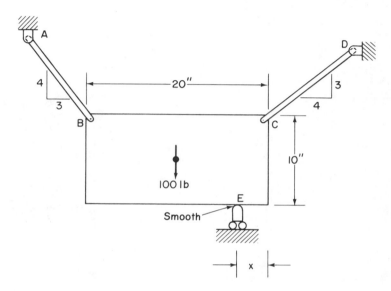

Figure P4-51

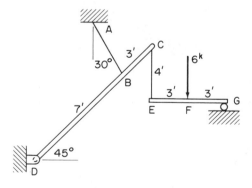

Figure P4-52

4-52. Determine the force in cable *AB* and the magnitude and direction of the force bar *CD* exerts on the pin at *D* in the structure shown in Fig. P4-52.

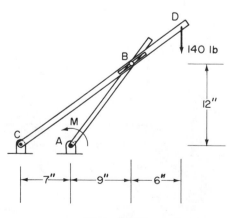

Figure P4-53

4-54. Find the force in *BD* of the structure shown in Fig. P4-54. Note that there are pins at *A*, *B*, *C*, *D*, and *E* and that *AD* and *BE* are not joined where they cross.

4-55. Figure P4-55 shows a hoist which travels along the track *DF*. The hoist *H* has a capacity of 2 kips, but the cable *BC* will break if the force in it exceeds 3 kips, and the cable *AB* will break if the force in it exceeds 4 kips. Find the maximum distance *x* that the hoist can be from *D* without either cable breaking.

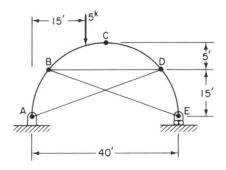

Figure P4-54

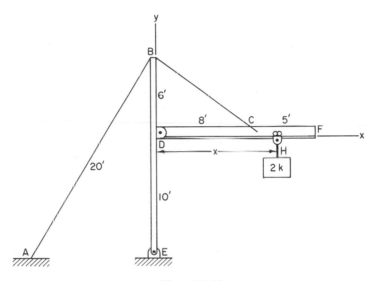

Figure P4-55

4-56. If the block *W* shown in Fig. P4-56 weighs 1000 lb, determine the force in *DB* and the *x* and *y* components of the reaction at *A*. The radius of the pulley at *C* is small enough that it may be neglected.

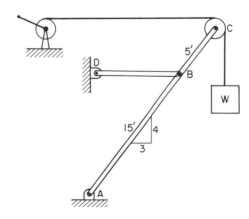

Figure P4-56

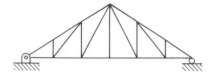

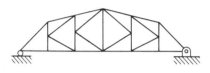

Figure 4-19

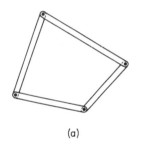

(a)

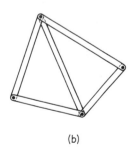

(b)

Figure 4-20

4-10. TRUSSES

A truss is a structure composed entirely of two force members connected in such a manner that they form a stable structure. The connections at each end of the members are treated as frictionless pinned joints, although the joints are usually riveted or, in modern structures, are frequently bolted or welded. Two of many possible styles or types of truss are shown in Fig. 4-19. Trusses were used very extensively in bridges and buildings in the late 1800's and early 1900's. They are not used as extensively in these applications today, but they are still used in other structures such as TV towers, truck frames, and supports for large roofs, such as gymnasiums. Trusses are useful in applications where a long span is required and the weight of the structure must be kept low.

The difference between a stable and an unstable structure made up of two-force members is shown in Fig. 4-20. The structure shown in Fig. 4-20(a) will deform, in fact it will collapse, if a load is applied to it. However, by adding another member to prevent deformation, the structure will act as a rigid body when a force is applied to it and will be able to support loads.

In the analysis of trusses, the objective is to find the force in each member of the truss so that the member can be made large enough to carry the load in it. There are two methods of analysis of forces in truss members. These are the *method of joints* and the *method of sections*. Which method is best depends on the particular problem, since some types of problems are solved more easily with one method than the other. Thus when you become familiar with both methods, you may select the most efficient method for solving a particular problem.

For many truss problems the first step is to find the reactions at the supports. Since a truss is treated as a rigid body, the reactions may be found by treating the truss as a beam. The difference is that it is a beam with varying depth. The example problems which follow will illustrate the method used to find the reactions at the supports.

4-11. METHOD OF JOINTS

The forces in each member of a truss can be analyzed by determining the forces acting on the pin at each joint.

These forces acting on the pin are caused by the forces in each of the members acting on the joint. A truss with a number of loads acting on it is shown in Fig. 4-21(a). The joint A on which the left reaction acts is shown isolated in Fig. 4-21(b). Note that, since all the members of the truss are two-force members, we will know the line of action of the force in each member. Therefore we can draw the free-body diagram of joint A as shown in Fig. 4-21(c). We note that each joint will be the point of concurrency for the forces in the members acting on it. Thus we will be limited to finding a maximum of two unknowns at any joint, since the equations available for our solution will be $\Sigma F_x = 0$ and $\Sigma F_y = 0$. For such a concurrent force system, a moment equation is usually an inefficient method of obtaining information.

   If we refer to the free-body diagram shown in Fig. 4-21(c), we can see that if we assume we have previously found the reaction at A by either analytical or graphical means, finding F_{AB} and F_{AH} should be a simple matter. If we sum forces in the y direction we have:

$$\Sigma F_y = 0$$
$$R_A - F_{AB} \sin \theta = 0$$
$$F_{AB} = \frac{R_A}{\sin \theta}$$

Similarly, now that F_{AB} has been obtained, we may find F_{AH} by summing forces in the x direction, as follows:

$$\Sigma F_x = 0$$
$$-F_{AH} - F_{AB} \cos \theta = 0$$
$$F_{AH} = -F_{AB} \cos \theta$$

The negative sign in the solution means that in drawing the free-body diagram we guessed wrong on the direction of the force F_{AH}. It is really opposite to the direction shown on the free-body diagram.

   Usually when solving a truss problem, it is a good idea to construct a table summarizing the results. Each member is listed along with the force in it. Tensile forces are shown with a positive sign, and compressive forces are shown with a negative sign. A *tensile force* is one which pulls on the joint, and a *compressive force* is one which pushes on the joint.

   Please note that the signs in the answer tables will not necessarily agree with those in the actual solution. The sign

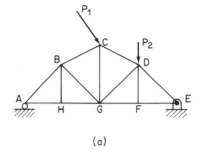

(a)

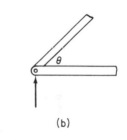

(b)

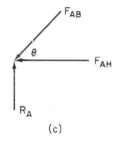

(c)

Figure 4-21

obtained in your solution simply tells you if you guessed right in drawing your free-body diagram. You must then decide from your free-body diagram whether the force is tension (a pull) or compression (a push). If all unknown forces in your free-body diagram are assumed to be tension, you will find that you automatically get the correct sign indicating force to be tension or compression. The forces in which you assumed the incorrect direction will always be compressive forces.

4-12. GRAPHICAL SOLUTION OF TRUSS PROBLEMS

Looking back at Fig. 4-21(c), you will see that the free-body diagram for the joint shows a concurrent force system. If you were to draw free-body diagrams for all the joints in the truss of Fig. 4-21(a), you would find that the forces on each joint constitute a concurrent force system. Unknown forces in such a system can quickly and conveniently be found by graphical methods in which we draw a force polygon. For equilibrium the force polygon must close. The method was discussed in Section 4-4.

The same procedure is used for obtaining the forces in trusses, with a minor modification to reduce the number of drawings from one per joint to just one per truss. The procedure, known as *Maxwell's method*, is illustrated in Fig. 4-22. This figure shows a truss with a load and the diagrams necessary to solve for the forces in each member of the truss. The Maxwell diagram is shown in Fig. 4-22(c). Although the diagram looks incomprehensible initially, it is really very simple when the principles used for its construction are understood. The Maxwell diagram is sometimes called a stress diagram. This is a most unfortunate name, since the diagram shows forces in members, not stresses.

Let us discuss the graphical procedure in an orderly fashion, beginning with the free-body diagram of the truss. The diagram in Fig. 4-22(b) is carefully drawn to scale. The spaces between the loads are labeled with letters using Bow's notation, and the spaces between members of the truss are numbered consecutively. Thus each member will be named by the number or letter on either side of it, such as *A*-1 or 1-2. The force **P** would be labeled *A*-*B* and the left reaction

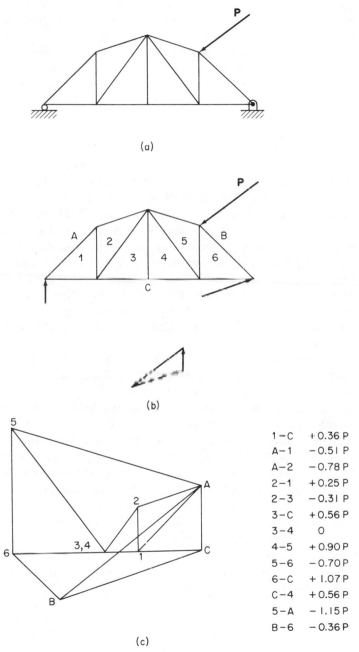

(a)

(b)

(c)

1−C	+0.36 P
A−1	−0.51 P
A−2	−0.78 P
2−1	+0.25 P
2−3	−0.31 P
3−C	+0.56 P
3−4	0
4−5	+0.90 P
5−6	−0.70 P
6−C	+1.07 P
C−4	+0.56 P
5−A	−1.15 P
B−6	−0.36 P

Figure 4-22

would be labeled *C-A*. In this case the free body was used to determine the reactions as shown in the small sketch at the bottom of Fig. 4-22(b). Note that the graphical solution for the reactions is repeated to a larger scale in Fig. 4-22(c).

Each joint in the truss is the point of concurrency for a concurrent force system. The Maxwell diagram is just the force polygons for all these concurrent force systems interconnected in a meaningful manner. To begin, pick a point where there are not more than two unknown forces. Usually the best starting point is one of the reactions, so in this case we start at the left reaction. The forces at that joint are read in sequence around the joint in a particular direction, either clockwise or counterclockwise. If read clockwise, they are *C-A*, *A*-1, and 1-*C*. For equilibrium they must form a closed triangle, which has been drawn in Fig. 4-22(c). The magnitude of *A-C* was known, and the directions of *A*-1 and 1-*C* were known, so drawing the vector triangle was no trouble. From this triangle the magnitudes of *A*-1 and 1-*C* can be scaled. Line *A*-1 in the Maxwell diagram is headed towards the joint; it would be in compression and is thus shown in the table with a negative sign. Similarly 1-*C* is headed away from the joint so it is in tension, or positive, as shown in the table of values.

Now we can consider the joint where 1-*A*, *A*-2, and 2-1 meet. Note that they are again labeled in clockwise sequence. Since 1-*A* is known and the lines of action of the other two forces are known, the vector triangle can again be readily solved. We proceed from joint to joint drawing the vector triangles, always moving on to a joint where there are no more than two unknown forces, and always proceeding in a fixed direction around the joint in labeling and drawing the vectors. Note that the Maxwell diagram should close, with the closing force being one of the reactions. You will also note that the points 3 and 4 on the Maxwell diagram coincide. This would infer that the force in member 3-4 is zero. That the force in 3-4 should be zero is fairly easy to understand if you draw a free-body diagram of the lower joint at the center of the truss and sum the forces in the y direction.

EXAMPLE 4-5

Determine the force in each member of the truss shown in Fig. 4-23(a).

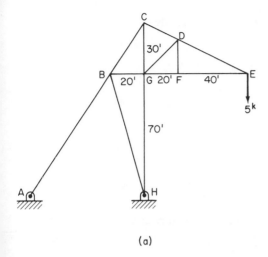

(a)

(b)

Figure 4-23

First let us solve the problem analytically. Draw the free-body diagram and solve for the reactions at A and H. Note that AB must be a two-force member, so that we know the line of action of the force in it. Using the slide rule method we can find the proportions of AB, since the proportions are the same as for BC. From this we can find the components of AB.

$$BC = 36.1 \text{ ft}$$

$$A_x = \frac{20}{36.1} A = .554A$$

$$A_y = \frac{30}{36.1} A = .831A$$

If we take moments about H, we can find the force A directly.

$$\Sigma M_H = 0$$

$$70 \times .554A + 20 \times .831A - (60 \times 5) = 0$$

$$38.8A + 16.6A - 300 = 0$$

$$55.4A - 300 = 0$$

$$A = \frac{300}{55.4} = 5.41 \text{ kips}$$

The x and y components of the force at H can be found by summing the forces in the x and y directions, respectively.

$$\Sigma F_x = 0$$

$$-.554A + H_x = 0$$

$$H_x = .554 \times 5.41 = 3.00 \text{ kips}$$

$$\Sigma F_y = 0$$

$$-.831A - 5.00 + H_y = 0$$

$$H_y = .831 \times 5.41 + 5.00 = 9.50 \text{ kips}$$

In order to find the forces in the members we must draw free-body diagrams of the joints so that the forces in them can be analyzed. Our solution should start at a point where there are not more than two unknown forces. Thus we could start at either joint E or H. For no particular reason, let us start the solution at joint E by drawing the appropriate free-body diagram, as shown in Fig. 4-24(a). Then sum forces in the y direction so that we will have one equation in one unknown.

From the free body of joint E

$$\Sigma F_y = 0$$

$$-5.00 + \frac{30}{67.1} F_{DE} = 0$$

$$F_{DE} = \frac{67.1}{30} \times 5.00 = 11.2 \text{ kips}$$

$$\Sigma F_x = 0$$

$$F_{FE} - \frac{60}{67.1} F_{DE} = 0$$

$$F_{FE} = \frac{60}{67.1} \times 11.2 = 10.0 \text{ kips}$$

From the free body of joint F

$$\Sigma F_x = 0$$

$$F_{FG} - F_{FE} = 0$$

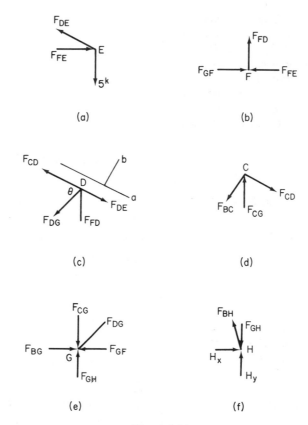

(a)

(b)

(c)

(d)

(e)

(f)

Figure 4-24

We have just shown that $F_{FD} = 0$. Thus at joint D, the solution can be made simpler if we draw our axes parallel to one of the unknown forces. Axes a and b are drawn in accordingly in Fig. 4-24(c).

$F_{FG} = 10.0$ kips

$\Sigma F_y = 0$

$F_{FD} = 0$

From the free body of joint D

$\Sigma F_b = 0$

$-F_{DG} \sin \theta = 0$

$F_{DG} = 0$

$\Sigma F_a = 0$

$F_{DE} - F_{CD} = 0$

$F_{CD} = 11.2$ kips

From the free body of joint C

$\Sigma F_x = 0$

$\dfrac{60}{67.1} F_{CD} - \dfrac{20}{36.1} F_{BC} = 0$

$F_{BC} = \dfrac{36.1}{20} \times \dfrac{60}{67.1} \times 11.2 = 18.1$ kips

$\Sigma F_y = 0$

$F_{CG} - \dfrac{30}{36.1} F_{BC} - \dfrac{30}{67.1} F_{CD} = 0$

$F_{CG} = \dfrac{30}{36.1} \times 18.1 + \dfrac{30}{67.1} \times 11.2$

$F_{CG} = 15.0 + 5.0 = 20.0$ kips

From the free body of joint G

$\Sigma F_y = 0$

$F_{GH} - F_{CG} = 0$

$F_{GH} = 20.0$ kips

$\Sigma F_x = 0$

$F_{BG} - F_{FG} = 0$

$F_{BG} = 10.0$ kips

From the free body of joint H

$\Sigma F_x = 0$

$H_x - \dfrac{20}{72.7} F_{BH} = 0$

$F_{BH} = 3.00 \times \dfrac{72.7}{20} = 10.9$ kips

It is a good idea to make a table of the forces in each member, so that all the results are together. Ordinarily, there is merit in making the table at the beginning of the solution so that you can conveniently refer back to previously found forces. Also, this way you are less likely to forget a force if you fill in the values in the table as you solve for them.

Member	Force (kips)
F_{BC}	+18.1
F_{CD}	+11.2
F_{DE}	+11.2
F_{GD}	0
F_{DF}	0
F_{EF}	−10.0
F_{FG}	−10.0
F_{CG}	−20.0
F_{BG}	−10.0
F_{BH}	+10.9
F_{GH}	−20.0

The graphical solution for the problem is shown in Fig. 4-25. The truss is first drawn to scale as shown in Fig. 4-25(a). The space between the forces are lettered using Bow's notation, and the spaces between the members are numbered. Since there are only

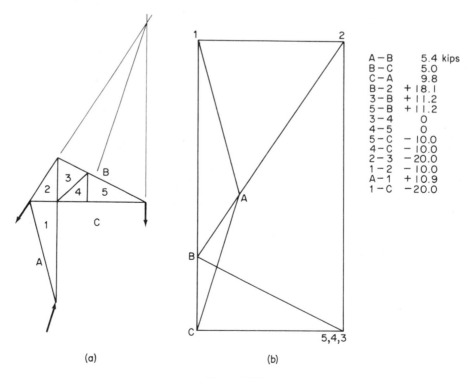

A − B	5.4	kips
B − C	5.0	
C − A	9.8	
B − 2	+ 18.1	
3 − B	+ 11.2	
5 − B	+ 11.2	
3 − 4	0	
4 − 5	0	
5 − C	− 10.0	
4 − C	− 10.0	
2 − 3	− 20.0	
1 − 2	− 10.0	
A − 1	+ 10.9	
1 − C	− 20.0	

(a)

(b)

Figure 4-25

three forces on the truss, Fig. 4-25(a) has also been used to determine the unknown line of action at the support *H* by determining the point of concurrency of the forces. The external forces are first solved for using the force polygon formed by *B-C*, *A-B*, and *A-C*. We then start solving for the forces in the members at any joint where there are not more than two unknown forces. The joint where the 5-kip load is applied is a good starting point. The force *B-C* is already drawn. If we proceed in a clockwise direction, we then draw in the forces for *C-5* and then *5-B*, each parallel to the member. This gives us a force triangle *C-B-5* representing the forces at the joint.

The forces can be scaled and their values placed in the table forming part of the graphical solution. Note that, by going in the clockwise direction, the names of the members are *C-5* and *5-B*. If we had arrows on the vectors, *C-5* would point towards the joint and thus is compression. Similarly, *5-B* points away from the joint and is thus in tension. We now consider the joint where *5-C*, *C-4*, and *4-5* meet. *5-C* is already drawn in our triangle. If we add *C-4* and *4-5* to get a closed triangle, we find that it can be done only if points 4 and 5 are common. This means that forces *5-C* and *C-4* must be the same and that force *4-5* must be zero. We can continue to the other joints in the same manner and find all the forces in the members. The total figure for the forces should close, since the truss as well as each member is supposed to be in equilibrium.

4-13. METHOD OF SECTIONS

This method of solving for forces in members of a truss is most useful if the forces in only a few isolated members are required. Frequently by this method it is possible to deter-

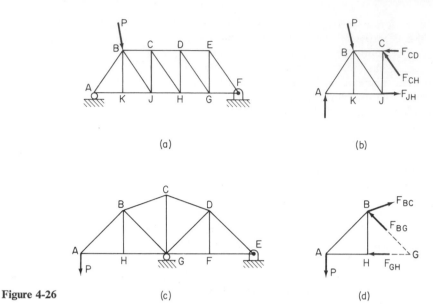

(a)

(b)

(c)

Figure 4-26

(d)

mine the force in a member near the middle of the truss in a single step, without having to find the forces in all the other members between it and the end. Two examples are shown in Fig. 4-26. Figure 4-26(a) shows a truss in which it is necessary to find the force in member *CH*. If a section is cut through members *CD*, *CH*, and *HJ*, and the cut members are replaced by the forces in them, we obtain a free-body diagram like that shown in Fig. 4-26(b). If the reaction at *A* has been found, then the force in *CH* could be found simply by summing forces in the *y* direction.

For the truss shown in Fig. 4-26(c) it is necessary to find the force in member *BC*. Again the truss is cut on a section which passes through members *BC*, *BG*, and *GH*, and the cut members are replaced by the forces in them, as shown in Fig. 4-26(d). The easiest way to find the force in *BC* would be to take moments about the point of concurrency of the other two forces, point *G*. This would permit finding the force in *BC* directly. Note that it is not necessary for the point about which moments are being taken to be on the free-body diagram.

In general, to use the method of sections, a section is passed through the truss cutting the member whose force is required. Note that since this is a plane equilibrium problem, we cannot solve for more than three unknowns. In general the section should not cut through more than three members whose forces are unknown, although there are a few special cases where partial solutions can be obtained if more than three members are cut.

EXAMPLE 4-6

Find the forces in members *DE*, *DN*, and *NO* of the truss shown in Fig. 4-27(a). There is a load of 2 kips at each of the joints on the lower chord between the supports.

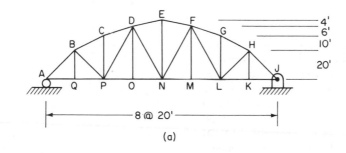

(a)

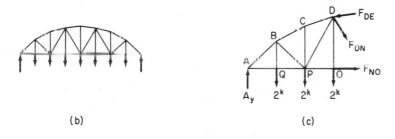

(b)   (c)

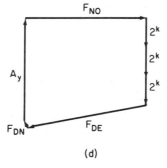

(d)

Figure 4-27

First make a free-body diagram of the truss to find the reactions. Since we plan to use the method of sections, we will need only one reaction. In this case we will find the left reaction, using the free-body diagram shown in Fig. 4-27(b). We could also find the reaction by using the fact that the structure is loaded symmetrically and that half the load must be carried by each support.

$\Sigma M_J = 0$

$20 \times 2 + 40 \times 2 + 60 \times 2 + 80 \times 2$
$\quad + 100 \times 2 + 120 \times 2 + 140 \times 2$
$\quad - 160 A_y = 0$

$160 A_y = 40 + 80 + 120 + 160 + 200$
$\quad\quad\quad\quad + 240 + 280$

$A_y = \dfrac{1120}{160} = 7.00 \text{ kips}$

In order to find the three forces required, we form a section which will cut through one or more of the members whose forces we seek. Here it is possible to have one section which will cut through all three members as shown in Fig. 4-27(c). Because of the orientation of the members, it is not possible to sum forces and get a single equation with only one unknown force. The best we can do is take moments about D so that we can find the force in NO.

$\Sigma M_D = 0$

$-(60 \times 7) + 40 \times 2 + 20 \times 2$
$\quad + 36 F_{NO} = 0$

$36 F_{NO} = 420 - 80 - 40$

$F_{NO} = \dfrac{300}{36} = 8.33 \text{ kips}$

The forces in both DE and DN have vertical and horizontal components so, if we summed forces, we would need to solve two equations simultaneously to find the forces in these members. Alternatively, we could take moments about any point on the line of action of either of the unknown forces. The easiest such solution would be to take moments about N, remembering that the line of action of the force in DE passes through point E.

$\Sigma M_N = 0$

$40 \times \left(\dfrac{20}{20.4}\right) F_{DE} + 20 \times 2 + 40 \times 2$
$\quad + 60 \times 2 - (80 \times 7.0) = 0$

$39.2 F_{DE} = -40 - 80 - 120 + 560$

$F_{DE} = \dfrac{320}{39.2} = 8.16 \text{ kips}$

The force in *DN* can now be found by summing forces in the *x* direction.

$$\sum F_x = 0$$

$$8.33 + \left(\frac{20}{41.2}\right) F_{DN} - \frac{20}{20.4} \times 8.16 = 0$$

$$F_{DN} = \frac{41.2}{20} \times (-8.33 + 8.00) = -0.67 \, \text{kips}$$

F_{DN} is 0.67 kips compression

Although the method of sections does not lend itself well to a purely graphical procedure, a check on results can be made graphically by using the results to draw a vector polygon. Since the truss and each part of the truss is in equilibrium, there is an error in the solution if the vector polygon does not close. But if the polygon does close, it does not guarantee that the solution is correct. Such a graphical check is shown in Fig. 4-27(d).

PROBLEMS

4-57. Find the tension in *BC*, the reaction at *D*, and the force in each member of the truss shown in Fig. P4-57.

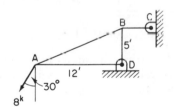

Figure P4-57

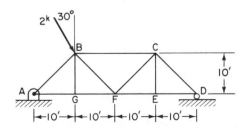

Figure P4-58

4-58. Determine the forces in members *AB*, *BC*, and *AG* of the truss shown in Fig. P4-58.

4-59. Determine the magnitude and kind of force in each member of the truss shown in Fig. P4-59. All triangles formed by the members are 45° triangles.

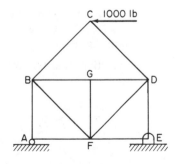

Figure P4-59

4-60. For the truss shown in Fig. P4-60, determine the force in each of the members.

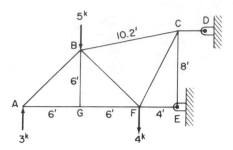

Figure P4-60

4-61. Find the force in each member of the truss shown in Fig. P4-61.

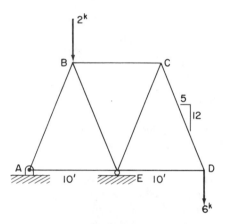

Figure P4-61

4-62. Using the method of joints, find the forces in members *DE*, *DG*, *CD*, and *CG* of the truss shown in Fig. P4-62.

4-63. Find the reactions and the forces in

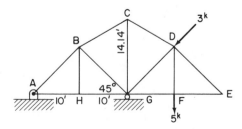

Figure P4-62

CG, *CH*, and *EH* of the truss shown in Fig. P4-63.

4-64. Find the reactions and the forces in the members *GH*, *CG*, and *GJ* of the truss shown in Fig. P4-64.

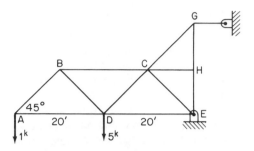

Figure P4-63

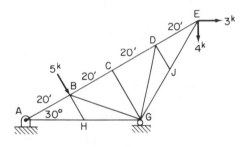

Figure P4-64

4-65. The truss shown in Fig. P4-65 is symmetrical. Find the forces in members *BG* and *CG*.

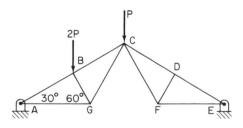

Figure P4-65

4-66. Determine the forces in members *BF*, *FG*, and *DH* of the truss shown in Fig. P4-66.

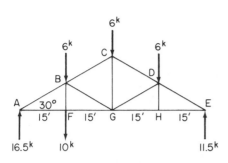

Figure P4-66

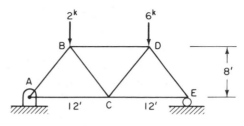

Figure P4-69

4-70. Determine the forces in members *AB* and *CD* of the truss shown in Fig. P4-70.

4-67. For the truss shown in Fig. P4-67, find the force in *AB* and the force in *DE*.

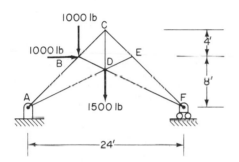

Figure P4-67

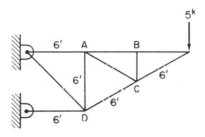

Figure P4-70

4-68. For the truss shown in Fig. P4-68, find the forces in *BF*, *BE*, and *CD*.

4-71. Find the components of the external reactions and find the forces in members *BC*, *BG*, and *AG* for the truss shown in Fig. P4-71.

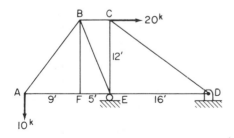

Figure P4-68

4-69. For the truss shown in Fig. P4-69, find the force in *BD*.

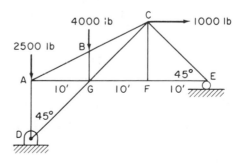

Figure P4-71

4-72. Find the force in member *AB* for the truss shown in Fig. P4-72.

4-73. Determine the force in member *CD* using the section *a-a* shown in Fig. P4-73.

4-74. Find the forces in members *CF* and *HG* for the truss shown in Fig. P4-74.

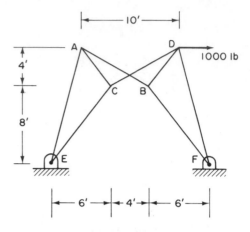

Figure P4-72

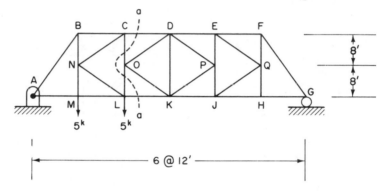

Figure P4-73

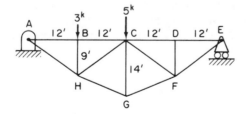

Figure P4-74

4-75. Find the forces in members *HD*, *HE*, and *AB* for the truss shown in Fig. P4-75.

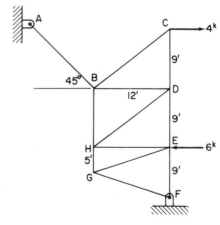

Figure P4-75

4-76. For the truss shown in Fig. P4-76, the force in member *BG* was found to be 30 kips tension. Find the magnitude of the force *P* applied to the truss, and find the force in *BH*.

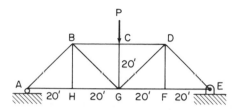

Figure P4-76

4-77. Given that the force in member *BE* in Fig. P4-77 is 12 kips compression, determine the load *P* on the truss.

4-78. If the load in member *EF* of the truss shown in Fig. P4-78 is 12 kips compression, determine the magnitude of the force *P*.

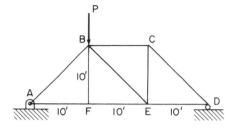

Figure P4-77

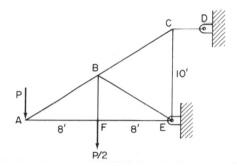

Figure P4-78

5

THREE-DIMENSIONAL STATICS

Until now the problems dealt with have been plane or two dimensional. Fortunately, although we live in a three-dimensional world, a large proportion of the problems which we encounter can be solved by a little manipulation—changing them to two-dimensional problems. However, sometimes it is either more convenient, or necessary, to solve some problems in three dimensions. You will find that the principles involved here are very similar to the principles involved in solving plane problems, although solutions may be a bit more time consuming because of the three dimensions.

A very large three-dimensional problem is shown in Fig. 5-1. However, you will discover that determining the forces in components of such a large structure is not very complicated if the procedures of this chapter are followed.

5-1. COMPONENTS AND DIRECTION COSINES

Figure 5-2 shows a force oriented in space. The rectangular parallelepiped shown helps to indicate the orientation of the force in space by showing the proportions of the force in the x, y, and z directions.

The axis system shown should always be used. The

Fig. 5-1. This 300 ton dockside crane is used for moving heavy pieces of equipment to and from ships. This is a three dimensional trussed structure, and can be analyzed using procedures from this and the previous chapter. (Photo courtesy of Toronto Harbour Commission.)

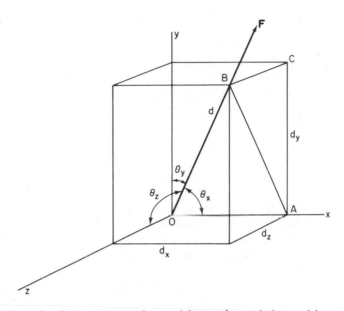

Figure 5-2

labeled ends of the axes are the positive ends, and the positive axes form what is called a right-hand triad. The thumb of the right hand points to the positive end of the x axis, the first finger points to the positive end of the y axis, and the middle finger points to the positive end of the z axis. As long as the axes have the same relative position as given by the right-hand triad, they can be oriented in any position in space.

The components of the force can be obtained by taking the product of the force and the direction cosine for the desired component. The orientation of the force is determined by measuring the angles called the direction angles between the positive x, y, and z axes and the force, as shown in Fig. 5-2. The x component of the force will be $F \cos \theta_x$. The truth of this statement will be more apparent if you consider the triangle formed by OAB. In this triangle, θ_x is the angle AOB, and the angle OAB is a right angle. Then the component of F in the x direction, using the right-angled triangle, is $F \cos \theta_x$. Similarly, then all the components may be expressed as:

$$F_x = F \cos \theta_x \qquad (5\text{-}1)$$

$$F_y = F \cos \theta_y \qquad (5\text{-}2)$$

$$F_z = F \cos \theta_z \qquad (5\text{-}3)$$

where

F_x, F_y, ánd F_z are the x, y, and z components respectively of the force **F**, and

θ_x, θ_y, and θ_z are the angles from the positive x, y, and z axes, respectively, and measured to the force **F**.

Instead of three angles, the numerical values of the direction cosines are frequently given, since the cosine is used more frequently than the angle itself. Note that the direction cosine can have a negative value. This has the same meaning that a negative cosine usually has, that is, the angle is greater than 90°.

The angles are not always given, in which case it is usually possible to obtain the direction cosines from the geometry of the problem. Referring to Fig. 5-2, the length of the diagonal of the parallelepiped, d, may be found in the following manner.

In the right-angled triangle ACB of Fig. 5-2, the length of AB is

$$d_{AB} = (d_y^2 + d_z^2)^{1/2}$$

In the right-angled triangle OAB

$$d_{OB} = d = (d_x^2 + d_{AB}^2)^{1/2}$$

but

$$d_{AB}^2 = d_y^2 + d_z^2$$

Thus

$$d = (d_x^2 + d_y^2 + d_z^2)^{1/2} \qquad (5\text{-}4)$$

where

d_x, d_y, and d_z are the x, y, and z components of the rectangular parallelepiped. Then the direction cosines will be

$$\cos \theta_x = \frac{d_x}{d} \qquad (5\text{-}5)$$

$$\cos \theta_y = \frac{d_y}{d} \qquad (5\text{-}6)$$

$$\cos \theta_z = \frac{d_z}{d} \qquad (5\text{-}7)$$

The values obtained for direction cosines may be checked by determining if the following relation is true for your values of direction cosines.

$$\cos^2 \theta_x + \cos^2 \theta_y + \cos^2 \theta_z = 1 \qquad (5\text{-}8)$$

The relationship shown in Eq. (5-8) is obtained in the following manner. Combining Eq. (5-5) through Eq. (5-8), we have

$$\cos^2 \theta_x + \cos^2 \theta_y + \cos^2 \theta_z = \left(\frac{d_x}{d}\right)^2 + \left(\frac{d_y}{d}\right)^2 + \left(\frac{d_z}{d}\right)^2$$

$$= \frac{d_x^2 + d_y^2 + d_z^2}{d^2}$$

but from Eq. (5-4)

$$d_x^2 + d_y^2 + d_z^2 = d^2$$

Thus

$$\cos^2 \theta_x + \cos^2 \theta_y + \cos^2 \theta_z = \frac{d^2}{d^2} = 1$$

EXAMPLE 5-1

Determine the x, y, and z components and the direction cosines for the 300-lb force shown in Fig. 5-3.

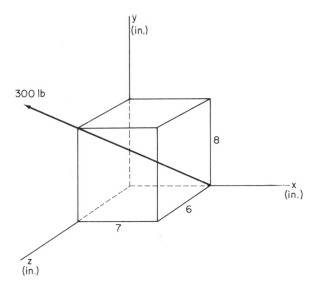

Figure 5-3

To calculate the components, we must first obtain the direction cosines. This can be done by using the lengths of the sides of the parallelepiped.

$$d = (d_x^2 + d_y^2 + d_z^2)^{1/2}$$
$$= (7^2 + 8^2 + 6^2)^{1/2}$$
$$= (49 + 64 + 36)^{1/2}$$
$$= (149)^{1/2}$$
$$= 12.2 \text{ in.}$$

The negative sign for $\cos \theta_x$ indicates that the angle between the positive x axis and $\mathbf{F}$ is larger than 90°.

$$\cos \theta_x = \frac{d_x}{d} = \frac{-7}{12.2} = -.574$$
$$\cos \theta_y = \frac{d_y}{d} = \frac{8}{12.2} = .656$$
$$\cos \theta_z = \frac{d_z}{d} = \frac{6}{12.2} = .491$$

The check is close enough to 1.00 to indicate that our calculations may be correct.

Check:
$$\cos^2 \theta_x + \cos^2 \theta_y + \cos^2 \theta_z$$
$$= (-.574)^2 + (.656)^2 + (.491)^2$$
$$= .330 + .430 + .241$$
$$= 1.001$$

Since the direction cosines are now obtained, we can calculate the components of the force.

$$F_x = F \cos \theta_x$$
$$= 300 \times (-.574)$$
$$= -172.4 \text{ lb}$$
$$F_y = F \cos \theta_y$$
$$= 300 \times .656$$
$$= 197.0 \text{ lb}$$
$$F_z = F \cos \theta_z$$
$$= 300 \times .491$$
$$= 147.2 \text{ lb}$$

PROBLEMS

5-1. Find the direction cosines for the force shown in Fig. P5-1.

5-2. Find the x, y, and z components of the force shown in Fig. P5-2.

5-3. The force shown is 87 lb. Find its x, y, and z components, using the dimensions of the parallelepiped shown in Fig. P5-3.

5-4. Find the direction cosines and the rectangular components of the 900-lb force shown in Fig. P5-4.

5-5. For the 500-lb force shown in Fig. P5-5, determine the rectangular components and θ_x, θ_y, and θ_z.

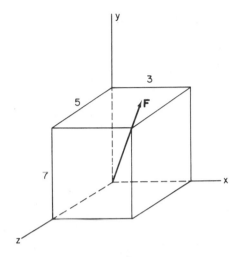

Figure P5-1

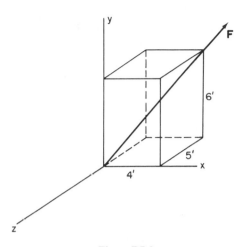

Figure P5-3

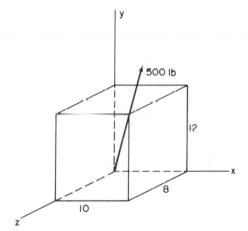

Figure P5-2

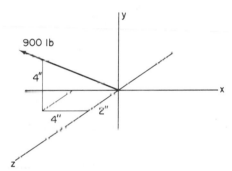

Figure P5-4

5-6. To provide an anchor for an antenna tower guy cable in sandy soil, it is sometimes necessary to form a large concrete anchor which weighs as much as the vertical component of the force in the cable. If a cable runs from the top of a tower to an anchor point which is 300 ft below the top of the tower, and 80 ft north and 60 ft east of the tower, and if the maximum force in the cable is 37 kips, determine the required weight for the anchor.

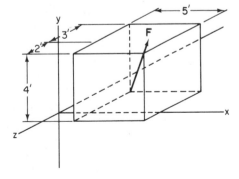

Figure P5-5

5-7. The boom for a quarry crane is shown in Fig. P5-7. If the force in the boom (a two-force member) is a maximum of 83 kips, determine the horizontal shear force that the connection at the base of the boom must resist. (The horizontal shear force is the resultant of the horizontal components of the force in the boom.)

5-8. A couple vector is shown in Fig. P5-8. Calculate the magnitude of the *x*, *y*, and *z* components and the direction cosines.

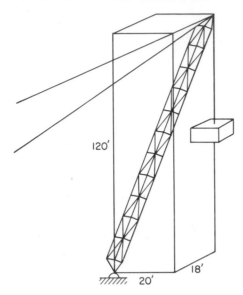

Figure P5-7

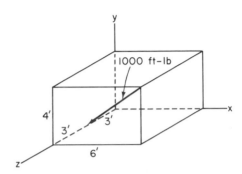

Figure P5-8

5-2. RESULTANTS FOR CONCURRENT FORCE SYSTEMS

In general, finding resultants in three dimensions is similar to finding them for a plane problem, except that our graphical methods are somewhat limited in three-dimensional problems. Whether solving graphically or analytically, it will be necessary to use components of the forces, summed in order to find resultants.

A few words of caution are in order here. Three-dimensional problems can become very complex. In this chapter we will deal primarily with resultants where the force systems are concurrent or parallel. However, a large number of nonconcurrent, nonparallel force problems can be solved without undue difficulty, although they do require a bit of time for their solution.

The resultant of several concurrent forces in three dimensions can be obtained by summing the components in the *x*, *y*, and *z* directions, so that we have

$$R_x = \sum F_{x_i} = F_{x_1} + F_{x_2} + F_{x_3} + \cdots \qquad (5\text{-}9)$$
$$R_y = \sum F_{y_i} = F_{y_1} + F_{y_2} + F_{y_3} + \cdots \qquad (5\text{-}10)$$
$$R_z = \sum F_{z_i} = F_{z_1} + F_{z_2} + F_{z_3} + \cdots \qquad (5\text{-}11)$$

The magnitude of the total resultant may then be found by a procedure similar to that used to obtain Eq. (5-4).

$$R = (R_x^2 + R_y^2 + R_z^2)^{1/2} \qquad (5\text{-}12)$$

where

R_x, R_y, and R_z are the x, y, and z components respectively of the resultant, and
R is the magnitude of the resultant.

The direction of the resultant is obtained by using the direction cosines. The numerical value of the direction cosine is usually found, so we do not bother finding the actual value of the angle unless there is a specific need for it.

$$\cos \theta_x = \frac{R_x}{R}$$

$$\cos \theta_y = \frac{R_y}{R}$$

$$\cos \theta_z = \frac{R_z}{R}$$

EXAMPLE 5-2

Find the resultant and its direction cosines for the two forces shown in Fig. 5-4. F_1 is 160 lb, and F_2 is 90 lb.

First let us find the lengths of the diagonals of the parallelepipeds so that the components of the forces can be calculated.

$$d_1 = (5^2 + 7^2 + 4^2)^{1/2}$$
$$= (25 + 49 + 16)^{1/2}$$
$$= (90)^{1/2}$$
$$= 9.49$$

$$d_2 = (4^2 + 5^2 + 3^2)^{1/2}$$
$$= (16 + 25 + 9)^{1/2}$$
$$= (50)^{1/2}$$
$$= 7.07$$

Now we can use the direction cosines to calculate the components.

$$F_{x_1} = 160 \times \frac{(-5)}{9.49} = -84.3 \text{ lb}$$

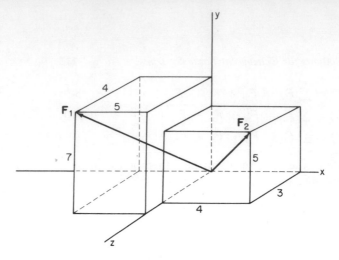

Figure 5-4

$$F_{y_1} = 160 \times \frac{7}{9.49} = 118.0 \, \text{lb}$$

$$F_{z_1} = 160 \times \frac{4}{9.49} = 67.4 \, \text{lb}$$

$$F_{x_2} = 90 \times \frac{4}{7.07} = 50.9 \, \text{lb}$$

$$F_{y_2} = 90 \times \frac{5}{7.07} = 63.6 \, \text{lb}$$

$$F_{z_2} = 90 \times \frac{3}{7.07} = 38.2 \, \text{lb}$$

The resultant can now be obtained by summing the components in the x, y, and z directions.

$$R_x = \Sigma \, F_{x_i} = -84.3 + 50.9 = -33.4 \, \text{lb}$$
$$R_y = \Sigma \, F_{y_i} = 118.0 + 63.6 = 181.6 \, \text{lb}$$
$$R_z = \Sigma \, F_{z_i} = 67.4 + 38.2 = 105.6 \, \text{lb}$$

The three-dimensional version of the Pythagorean theorem is used to find the magnitude of the resultant of the three components.

$$R = (33.4^2 + 181.6^2 + 105.6^2)^{1/2}$$
$$= (1114 + 33000 + 11120)^{1/2}$$
$$= (45200)^{1/2}$$
$$= 213 \, \text{lb}$$

The direction cosines are:

$$\cos \theta_x = \frac{-33.4}{213} = -.1565$$

$$\cos \theta_y = \frac{181.6}{213} = .852$$

$$\cos \theta_z = \frac{105.6}{213} = .496$$

This answer is close enough to 1.00 to indicate that we probably have no serious error in our solution.

Check:

$$\cos^2 \theta_x + \cos^2 \theta_y + \cos^2 \theta_z$$
$$= (-.1565)^2 + (.852)^2 + (.496)^2$$
$$= .0245 + .726 + .246 = .997$$

PROBLEMS

5-9. Find the resultant of the forces shown in Fig. P5-9 and the direction cosines of the resultant.

5-10. Find the resultant of the two forces shown in Fig. P5-10.

5-11. Find the magnitude and direction of the resultant of the two forces shown in Fig. P5-11. The 80-lb force is in the *x-y* plane.

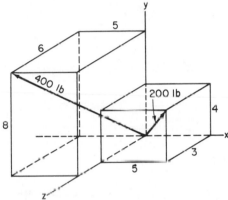

Figuro P5-9

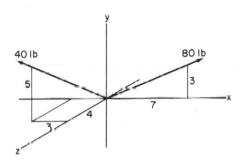

Figure P5-11

5-12. Determine the magnitude of the resultant of the forces shown in Fig. P5-12.

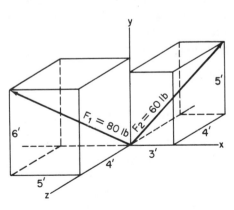

Figure P5-10

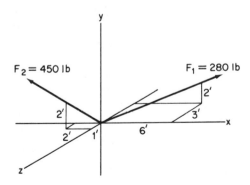

Figure P5-12

5-13. In order to check the size of the anchor at *A* in Fig. P5-13, it is necessary to know the resultant force from the two cables *AB* and *AC*. Determine the resultant, if the force in *AB* is 280 lb, and the force in *AC* is 115 lb.

5-14. Find the magnitude and direction of the resultant of the three forces shown in Fig. P5-14.

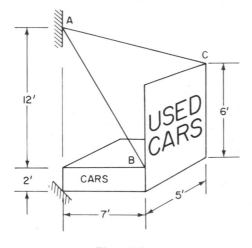

Figure P5-13

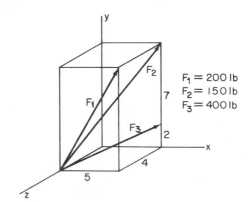

$F_1 = 200$ lb
$F_2 = 150$ lb
$F_3 = 400$ lb

Figure P5-14

5-3. GRAPHICAL PROCEDURES

Three-dimensional problems can be solved using graphical procedures. However, there is some question about the value of discussing them in this book, for they involve the use of a modest amount of descriptive geometry in order to find true line lengths and true angles. If you are interested in the graphical approach to solving three-dimensional problems in statics, you should refer to a book on descriptive geometry or engineering drawing which discusses three-dimensional statics problems.

5-4. RESULTANT OF PARALLEL FORCES

The resultant for a system of parallel forces such as shown in Fig. 5-5 is quite easy to find, for as one might expect, the magnitude of the resultant is simply the algebraic sum of the forces, as follows:

$$R = \sum F_i = F_1 + F_2 + F_3 + \cdots \qquad (5\text{-}13)$$

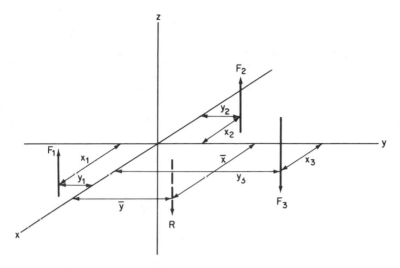

However, the resultant must have the same effect on a body as the initial system of forces. This means that the resultant must also cause the same moment about any point as the initial system of forces. Generally the moments are taken about convenient axes, usually the x and y axes, instead of being taken about a point. Then if $\bar{x}$ and $\bar{y}$ are the x and y coordinates of the resultant, their value may be obtained by setting the moment of the resultant equal to the moment of all the forces, as follows:

$$R\bar{x} = \Sigma (F_i x_i) = F_1 x_1 + F_2 x_2 + F_3 x_3 + \cdots$$
$$\bar{x} = \frac{\Sigma (F_i x_i)}{R} \qquad (5\text{-}14)$$

and

$$R\bar{y} = \Sigma (F_i y_i) = F_1 y_1 + F_2 y_2 + F_3 y_3 + \cdots$$
$$\bar{y} = \frac{\Sigma (F_i y_i)}{R} \qquad (5\text{-}15)$$

where
  F_1, F_2, and F_3 etc. are the magnitudes of the given forces.
  x_1, x_2, and x_3 etc., and y_1, y_2, and y_3 etc. are the x and y coordinates of the points where the the forces pass through the x-y plane,
  R is the magnitude of the resultant of the forces,
  $\bar{x}$ is the x coordinate of the resultant force, and
  $\bar{y}$ is the y coordinate of the resultant force.

Figure 5-5

If a positive sign is obtained for our values of $\bar{x}$ or $\bar{y}$, this indicates that the assumed location of the resultant was correct. If a negative sign is obtained, it means that for this co-ordinate its location should be on the opposite side of the axis to that assumed.

EXAMPLE 5-3

Replace the system of forces shown in Fig. 5-6(a) with a single force and determine where the single force should be located.

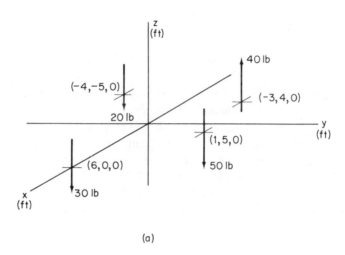

(a)

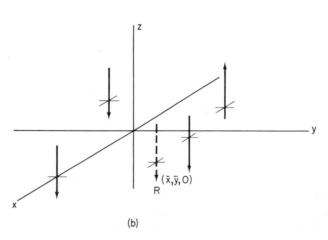

(b)

Figure 5-6

First we should find the magnitude of the resultant by summing the forces.

$$R = \Sigma F_i$$
$$= -20 - 30 - 50 + 40$$
$$= -60 \text{ lb}$$

Next, to find the x coordinate of the point through which the force must pass, we take moments about the y axis. The assumed location of the resultant is shown in Fig. 5-6(b).

$$\Sigma M_y:$$
$$R\bar{x} = \Sigma (F_i x_i)$$
$$60\bar{x} = -(4 \times 20) + 3 \times 40 + 1 \times 50 + 6 \times 30$$
$$= -80 + 120 + 50 + 180$$
$$= 270$$
$$\bar{x} = \frac{270}{60} = 4.50 \text{ ft}$$

Similarly, to find the y coordinate of the point through which the resultant passes, we take moments about the x axis. The sign associated with the moments is that obtained from the right-hand rule.

$$\Sigma M_x:$$
$$R\bar{y} = \Sigma (F_i y_i)$$
$$-(60\bar{y}) = 5 \times 20 + 4 \times 40 - (5 \times 50) + 0 \times 30$$
$$= 100 + 160 - 250$$
$$= 10$$

Since we obtained a negative value for $\bar{y}$, the point must be located on the opposite end of the axis to that assumed.

$$\bar{y} = \frac{10}{-60} = -.167 \text{ ft}$$

The line of action of the resultant passes through the point with coordinates of $(4.50, -167, 0)$ ft.

PROBLEMS

5-15. Find and locate the resultant of the three forces shown in Fig. P5-15.

5-16. When designing a foundation for heavy machinery, it is necessary to locate the resultant of the several forces the machine exerts on the foundation. Figure P5-16 shows the forces exerted by a machine. Locate the resultant of these forces.

5-17. Find and locate the resultant of the four forces shown in Fig. P5-17.

5-18. Find and locate the resultant for the five forces shown in Fig. P5-18.

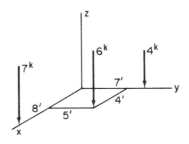

Figure P5-15

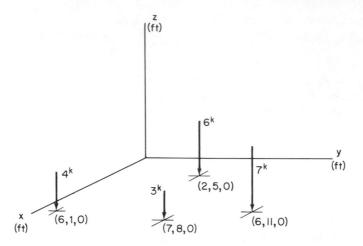

Figure P5-16

5-19. A decorative mobile for an art gallery is made up of three smaller mobiles weighing 2, 7, and 4 lb which are suspended from an 8-lb plate as shown in Fig. P5-19. In order to support the mobile properly, the support must act through the same point as the resultant of the four forces. Locate the resultant. The weight of the plate may be assumed to act through its center.

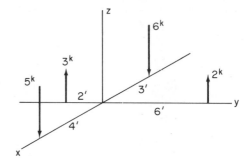

Figure P5-17

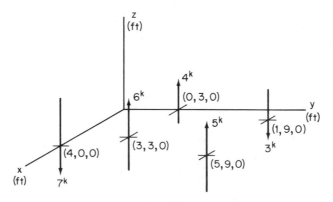

Figure P5-18

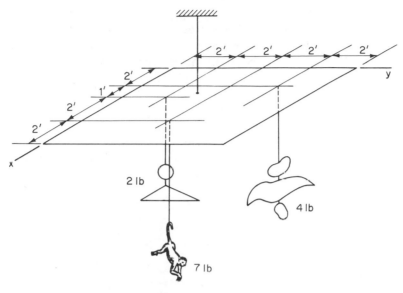

Figure P5-19

5-20. The platform (assumed weightless) shown in Fig. P5-20 has drums *A*, *B*, and *C* on it, weighing 5, 4, and 3 kips, respectively. Where should a fourth drum, weighing 6 kips, be placed so that the resultant force acts through the point (0, 0, 0)?

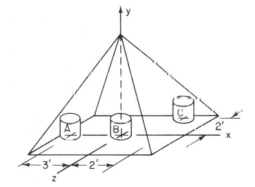

Figure P5-20

5-5. EQUILIBRIUM IN THREE DIMENSIONS

It is frequently necessary to find the forces required to maintain a body in equilibrium in three-dimensional systems. Although equilibrium problems can be very complex, we will restrict our discussion to the more elementary cases of concurrent force systems and simple nonconcurrent force systems.

The rules for equilibrium in three dimensions are similar to the rules for two dimensions described in Chapter 4.

For a body to be in equilibrium, two conditions must exist—the sum of all the forces acting on the body must be zero, and the sum of the moments of all the forces about any point must also be zero. This is expressed as:

$$\Sigma \mathbf{F} = 0 \tag{4-1}$$

$$\Sigma \mathbf{M} = 0 \tag{4-2}$$

Usually these expressions will be given in terms of the x, y, and z components of the forces, and the moments about the x, y, and z axes, so that we have

$$\Sigma F_x = 0$$
$$\Sigma F_y = 0$$
$$\Sigma F_z = 0$$
$$\Sigma M_x = 0$$
$$\Sigma M_y = 0$$
$$\Sigma M_z = 0$$

With the six equations above, it would appear that it is possible to solve for six unknowns, which could be in the form of forces, angles, or moments. This is true, but in the case of a concurrent force system, only the three equations dealing with summing forces are useful. The moment equations are an inefficient means of obtaining useful information when we impose the restriction that the forces be concurrent.

If the forces are all parallel to one axis, we can solve for only three unknowns; by imposing this condition, we in a sense use up three of the equations—the summation of the forces in the two directions in which there are no forces, and one moment equation.

For more general types of three-dimensional problems, up to the total of six equations can be used to solve for up to six unknowns. Regardless of the number of unknowns, a free-body diagram will be a great help in setting up your problems for solution. As usual, care must be taken to show all forces on the free-body diagram. Generally, it is best to show the rectangular components of the forces. When writing equilibrium equations to solve for unknown forces, it is usually a good idea to try to make the equations as simple as possible so that they contain few unknowns. However, sometimes you can spend too much time looking for an easy solution, so that it may be preferable to set up your equilibrium equations and spend your time carefully solving systems of simultaneous equations or using determinants for the solution of your equations.

More than six unknown quantities in your free-body diagram results in a statically indeterminate problem which cannot be solved using the methods of statics only.

The following two example problems illustrate the use of the equations of equilibrium to solve three-dimensional problems.

EXAMPLE 5-4

Determine the tension in each of the three cables which support the 200-lb weight shown in Fig. 5-7(a).

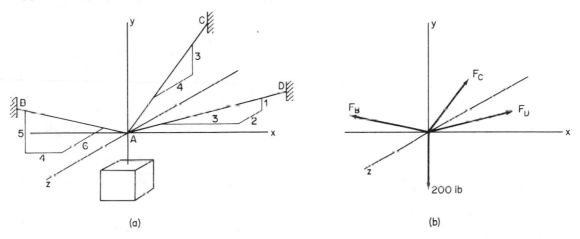

(a)   (b)

Figure 5-7

It will be necessary to find the components of the force in each cable in order to use the equilibrium equations, so we first need to find the proportion of each cable in each direction. To do this we calculate the diagonal length for each of the figures showing the orientation of the cables.

$$d_B = (4^2 + 5^2 + 6^2)^{1/2}$$
$$= (16 + 25 + 36)^{1/2}$$
$$= (77)^{1/2} = 8.77$$

$$d_C = (3^2 + 4^2)^{1/2}$$
$$= (9 + 16)^{1/2}$$
$$= (25)^{1/2} = 5.00$$

$$d_D = (3^2 + 1^2 + 2^2)^{1/2}$$
$$= (9 + 1 + 4)^{1/2}$$
$$= (14)^{1/2} = 3.74$$

The free-body diagram of the point where the three cables and the rope supporting the 200-lb weight join is shown in Fig.

$$\Sigma F_x = 0$$

$$\left(\frac{-4}{8.77}\right)F_B + \left(\frac{3}{3.74}\right)F_D = 0 \qquad \text{(a)}$$

5-7(b). For equilibrium the sum of the forces in each direction must be zero.

$\Sigma F_y = 0$

$$\left(\frac{5}{8.77}\right) F_B + \frac{3}{5} F_C + \left(\frac{1}{3.74}\right) F_D - 200 = 0$$
(b)

$\Sigma F_z = 0$

$$\left(\frac{6}{8.77}\right) F_B - \left(\frac{4}{5}\right) F_C - \left(\frac{2}{3.74}\right) F_D = 0 \quad \text{(c)}$$

We now have three simultaneous equations which we can solve for F_B, F_C, and F_D.

From Eq. (a)

$$F_B = \frac{8.77}{4} \times \left(\frac{3}{3.74}\right) F_D = 1.76 F_D$$

Substituting for F_B in Eq. (b)

$$\frac{5}{8.77} \times 1.76 F_D + \left(\frac{3}{5}\right) F_C + \left(\frac{1}{3.74}\right) F_D$$

$$-200 = 0$$

$$1.00 F_D + .60 F_C + .268 F_D - 200 = 0$$

$$.60 F_C + 1.268 F_D - 200 = 0 \qquad \text{(d)}$$

Substituting for F_B in Eq. (c)

$$\frac{6}{8.77} \times 1.76 F_D - \left(\frac{4}{5}\right) F_C - \left(\frac{2}{3.74}\right) F_D = 0$$

$$1.20 F_D - .80 F_C - .535 F_D = 0$$

$$-.80 F_C + .665 F_D = 0 \qquad \text{(e)}$$

From Eq. (e)

$$F_C = \left(\frac{.665}{.80}\right) F_D = .831 F_D$$

Substituting for F_C in Eq. (d)

$$.60 \times .831 F_D + 1.268 F_D - 200 = 0$$

$$.499 F_D + 1.268 F_D = 200$$

$$1.767 F_D = 200$$

$$F_D = \frac{200}{1.767} = 113 \text{ lb}$$

Substituting for F_D in Eq. (e) as rewritten

$$F_C = .831 \times 113 = 94.2 \text{ lb}$$

Substituting for F_D in Eq. (a) as rewritten

$$F_B = 1.76 \times 113 = 199 \text{ lb}$$

EXAMPLE 5-5

An idler shaft and pulley are shown in Fig. 5-8(a). Determine the x and y components of the reactions at bearings A and B. Due to friction in the bearings, there are torques in both A and B which resist rotation. Assume that the torques are equal in both bearings.

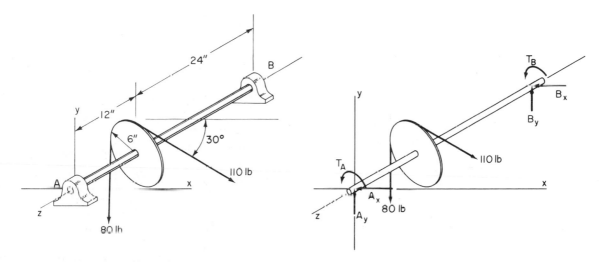

Figure 5-8

The first step is to draw the free-body diagram as shown in Fig. 5-8(b). If we count the unknowns on our free-body diagram, we note that there are six. If we are careful in the way we set up our solution, we discover that the problem can be solved without using simultaneous equations.

If we consider all the possibilities, we see that if we take moments about the z axis we will have the unknowns T_A and T_B in our equation. But since $T_A = T_B$, this moment equation can quickly be reduced to an equation with only one unknown.

$$\sum M_z = 0$$

$$-(110 \times 6) + 80 \times 6 + T_A + T_B = 0$$

$$-660 + 480 + T_A + T_B = 0$$

$$T_A + T_B = 180$$

but

$$T_A = T_B$$

thus

$$2T_A = 180$$

$$T_A = \frac{180}{2} = 90 \text{ in.-lb}$$

$$T_B = 90 \text{ in.-lb}$$

If we take moments about the x axis, we will have another equation with only one unknown.

$$\sum M_x = 0$$

$$36B_y - (80 \times 12)$$
$$- (110 \times \sin 30° \times 12) = 0$$

$$36B_y - 960 - 660 = 0$$

$$B_y = \frac{1620}{36} = 45 \text{ lb}$$

Now A_y is the only unknown force in the y direction.

$$\sum F_y = 0$$

$$A_y - 80 - 110 \sin 30° + B_y = 0$$

$$A_y = 80 + 55 - 45 = 90 \text{ lb}$$

The easiest way to find B_x is to take moments about the y axis.

$$\sum M_y = 0$$

$$36B_x - 110 \cos 30° \times 12 = 0$$

$$B_x = \frac{1142}{36} = 31.8 \text{ lb}$$

The last unknown A_x can now be found by summing forces in the x direction.

$$\sum F_x = 0$$

$$-A_x + 110 \cos 30° - B_x = 0$$

$$A_x = 95.3 - 31.8 = 63.5 \text{ lb}$$

PROBLEMS

5-21. The three ropes shown in Fig. P5-21 support a 500-lb box. What is the tension in each rope?

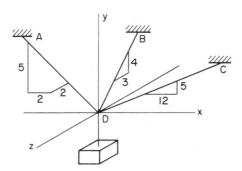

Figure P5-21

5-22. A 500-lb weight shown in Fig. P5-22 is supported by three cords each 10 ft long which are attached to points equally spaced around a ring which is 12 ft in diameter. Find the tension in each cord.

5-23. The 6 ft × 8 ft plate shown in Fig. P5-23 weighs 1000 lb. Determine the force in each cable, if point D is on the y axis.

5-24. Determine the force in each leg of the tripod shown in Fig. P5-24. The 1000 lb force is parallel to the x axis.

5-25. Find, in terms of W, the tension in each cable shown in Fig. P5-25. W is a weight hanging from point O. Points A, B, and C are fixed in the positions shown.

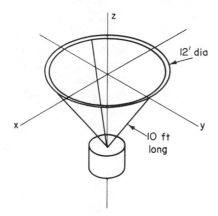

Figure P5-22

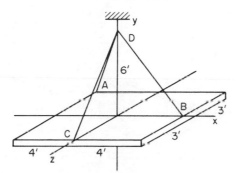

Figure P5-23

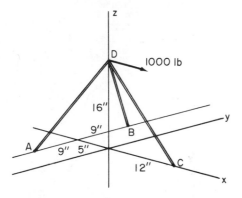

Figure P5-24

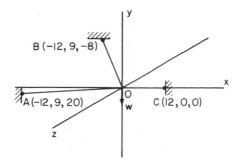

Figure P5-25

5-26. The two-force member space structure shown in Fig. P5-26 supports a vertical load F of 20 kips. The coordinate points are given in feet. Find the tension in the cable CD. A and B are ball and socket joints.

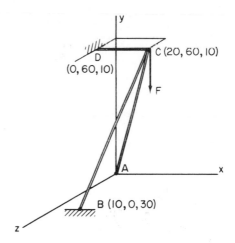

Figure P5-26

5-27. Figure P5-27 shows a tripod carrying a load P. Coordinates of the ends of the tripod legs are given in feet. If no leg can carry more than 100 lb, calculate the maximum load P.

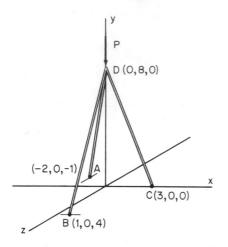

Figure P5-27

5-28. Once upon a time a crank was standard equipment on a car. It was turned in order to start the car. How much moment does the 100-lb cranking force provide about axis *a-a* shown in Fig. P5-28?

5-29. A 6 ft × 4 ft sign weighing 150 lb is supported by a ball and socket at *D* and two cables, as shown in Fig. P5-29. Determine the tension in the cables.

5-30. Determine the force *P* in the *y* direction and the bearing reactions required to maintain equilibrium for the axle shown in Fig. P5-30. All centerlines are in the *x-z* plane. The bearings at *A* and *B* are smooth and support no thrust loads.

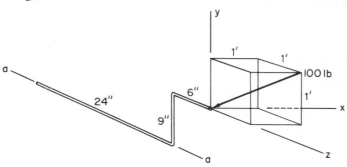

Figure P5-28

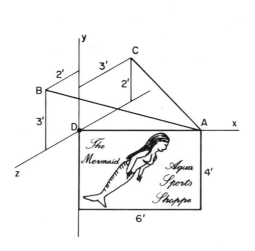

Figure P5-29

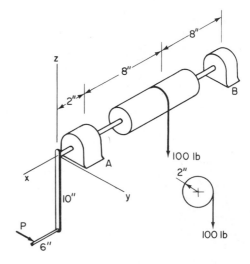

Figure P5-30

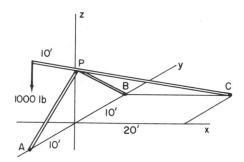

Figure P5-31

5-31. Determine the components of the reactions at the points *A*, *B*, and *C* for the space frame shown in Fig. P5-31. Point *P* is 15 ft above the *x-y* plane.

5-32. Replace the couple and two forces shown in Fig. P5-32 by a resultant force **R** acting at point *A* and a couple **M**.

5-33. The wall frame shown in Fig. P5-33 is loaded with a 6-kip force in the *y* direction at the center of bar *AB*. A 2-kip force is applied in the *z* direction at *B*. Members *CB* and *EB* are struts which can be loaded in tension or compression. Determine (a) the *x*, *y*, and *z* components of the reaction at *A*, and (b) the force in member *CB*.

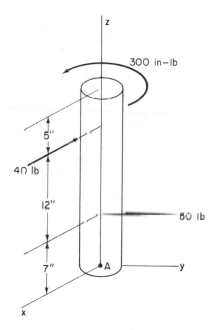

Figure P5-32

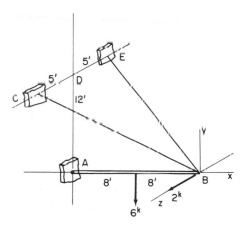

Figure P5-33

6

FRICTION

The phenomenon known as friction is a scientific problem about which we know very little, even though we like to think that we now have a good knowledge of why things behave the way they do. We recognize that friction is a source of both help and inconvenience, for in most of our daily life we have to fight to overcome friction. Yet we also find that we can not get anywhere without the benefit of friction. The automobile, as shown in Fig. 6-1, is a good example of the problems involved. Without friction there would be no traction between the tires and the pavement, and hence no motion. On the other hand, much money and effort is spent in overcoming the wear caused by friction in the engine.

6-1. CAUSES OF FRICTION

A friction force will exist between any two bodies in contact which move or try to move relative to one another. This force will exist regardless of how smooth the surfaces are, even if one of the surfaces is a fluid.

The mechanism of friction may be considered as some combination of two different physical phenomenon. First, regardless of how carefully machined and polished a surface is,

Fig. 6-1. Drag racing enthusiasts have some interesting problems with friction. Maximum friction is required between the slicks and the pavement, but minimum friction is required in the engine and transmission so that as little power as possible is wasted in overcoming friction. (Photo courtesy of Hot Rod Magazine.)

no surface is perfectly smooth, so that under sufficient magnification it may appear as shown in Fig. 6-2. As part *A* tries to move relative to part *B*, the jagged edges tend to interfere and lock, thus preventing motion. If sufficient force is applied, the two bodies can move, one relative to the other, either by breaking off some of the peaks or by riding up over the resisting projections.

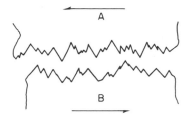

Figure 6-2

The other mechanism which may contribute to friction is the attractive forces that exist between molecules. The molecules of the two bodies may be very close together in carefully polished parts, and the sum of the many molecular attractive forces may be enough to tend to resist motion.

Our study of friction will be restricted to dry or Coulomb friction. (C. A. de Coulomb, the French physicist, performed some of the earliest experiments on friction and stated some laws of friction.) The frictional forces occurring if a lubricant is applied between the surfaces will not be included in our discussion.

Despite considerable study since Coulomb's time most of our laws of friction are still based on empirical or experimental evidence.

6-2. LAWS OF FRICTION

The laws of friction listed below have been developed experimentally, and some of them are generalizations which are not always precisely true. In other words, minor deviations from some of the rules will be found. Nonetheless, for want of better rules, we will accept those listed as valid.

1. The friction force always opposes relative motion.
2. The friction force is parallel to or tangent to the contact surfaces.
3. The friction force is independent of the area of contact.
4. For impending motion and motion at a uniform speed, the friction force is equal to the product of the coefficient of friction and the normal force. (Impending motion is motion which is just about to begin.)

Of the laws listed above, (1) and (2) are always valid.

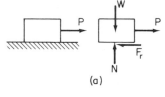

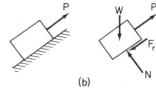

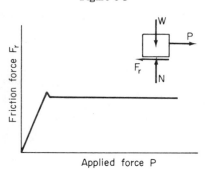

6-3. THE COEFFICIENT OF FRICTION

For any body resting on a surface, there will be a normal force. This force is perpendicular to the surface and is usually caused by the surface pushing up or supporting the body. The friction force, being parallel to the surface, will thus be perpendicular to the normal force. Two examples of these forces are shown in Fig. 6-3.

The relationship between the normal force and the friction force is

$$\mu = \frac{F_r}{N} \qquad (6\text{-}1)$$

where

μ is the coefficient of friction,
F_r is the friction force, and
N is the normal force.

This relationship is applicable only if motion is impending or if motion is at a uniform speed. Usually the coefficient of friction for impending motion, known as the static coefficient of friction, μ_s, is higher than the kinetic coefficient of friction for motion, which is μ_k. Figure 6-4 shows how the frictional force varies with impending motion and eventual motion at low speeds. (For most of our discussion, we will use μ for the coefficient of friction, without using the subscript.)

Values for the coefficient of friction are obtained by experimentation. Some typical values are shown in Table 6-1. You will note that a range of values is given. This is due to the wide variety of surface conditions that are encountered, either in terms of smoothness or degree of lubrication, since lubricants are never completely absent.

Table 6-1. Static Coefficients of Friction

Materials	μ_s
Wood and metal	0.2–0.3
Leather and wood	0.3–0.4
Asbestos compound on metal	0.3–0.4
Rubber tires on smooth dry pavement	0.7–0.9
Metal on ice	0.02–0.04

6-4. DETERMINATION OF THE COEFFICIENT OF FRICTION

One way to obtain the static coefficient of friction is by means of the inclined plane as shown in Fig. 6-5. The block is composed of one of the materials and the plane of the other material for which the coefficient of friction is desired.

The plane is raised very gradually until the block just starts to slide, and the angle of inclination θ is noted. When motion impends (that is, when motion is just about to begin), the component of the weight of the block parallel to the plane is just sufficient to balance the maximum friction force as shown in Fig. 6-5(b). The component of the weight parallel to the plane is $W \sin \theta$. The component of the weight normal to the plane is $W \cos \theta$. For equilibrium, which must exist if motion impends, the following relationships must be true:

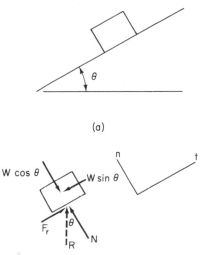

(a)

(b)

Figure 6-5

$$\Sigma F_n = 0$$
$$N - W \cos \theta = 0$$
$$N = W \cos \theta$$
$$\Sigma F_t = 0$$
$$F_r - W \sin \theta = 0$$
$$F_r = W \sin \theta$$

but by definition

$$\mu = \frac{F_r}{N} = \frac{W \sin \theta}{W \cos \theta} = \frac{\sin \theta}{\cos \theta} \qquad (6\text{-}2)$$

If we refer to the article on trigonometric functions in Chapter 2, we see that $\sin \theta / \cos \theta = \tan \theta$. Thus

$$\mu = \frac{\sin \theta}{\cos \theta} = \tan \theta$$

We conclude that the coefficient of static friction is also equal to the tangent of the angle at which motion would impend down an inclined plane. This angle is known as the *angle of repose*. It is also the angle between the normal force and the resultant of the friction force and the normal force, as shown in Fig. 6-5(b), when motion is impending.

Note that if we check the values for tangents, there is really no limitation on the value of the coefficient of friction. If two rough materials are used, one as an inclined plane and one as a block, it is possible to find combinations for which the angle θ will be more than 45°, and consequently μ will be more than 1.00.

The angle of repose is useful in ways other than determining the angle at which a block will just start to slide down a plane. The angle of a pile of coal, gravel, or other granular material depends on the moisture content and the friction between particles which varies from material to material. The angle of repose for the material will govern how steeply it may be piled, influencing how large an area of ground is required to store a given quantity of material. Figure 6-6 illustrates the angle of repose for a pile of granular material.

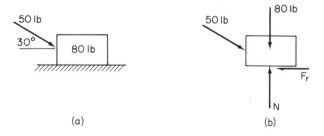

Figure 6-6

EXAMPLE 6-1

The 80-lb block shown in Fig. 6-7(a) is just about to move with the 50-lb force applied. Determine the coefficient of static friction.

(a)   (b)

Figure 6-7

Most friction problems are really just equilibrium problems with a different type of force in the problem than we encountered previously. Since the problem is an equilibrium problem the first step is to draw a free-body diagram as shown in Fig. 6-7(b).

We can sum forces in the y direction in order to find the normal force. Make sure that you include the y components of all forces.

$$\Sigma F_y = 0$$

$$N - 50 \sin 30° - 80 = 0$$

$$N = 25 + 80$$

$$= 105 \text{ lb}$$

Now the friction force can be obtained by summing forces in the x direction.

$$\Sigma F_x = 0$$

$$50 \cos 30° - F_r = 0$$

$$F_r = 43.3 \text{ lb}$$

$$\mu = \frac{F_r}{N} = \frac{43.3}{105}$$

$$= .412$$

Since friction problems are primarily equilibrium problems, they may be solved or checked graphically in many cases. The graphical solution is shown in Fig. 6-8. The two given forces, the weight and the applied force, are drawn to scale. Then the normal and friction forces are drawn in their proper directions, forming a closed polygon. The values of F_r and N may be scaled from the polygon and used to calculate μ, or the resultant of F_r and N, as shown by the dashed line, may be drawn. The angle between the resultant and N can be used to obtain the coefficient of friction, since $\tan \theta = \mu$ when motion impends.

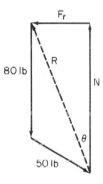

Figure 6-8

EXAMPLE 6-2

The block shown in Fig. 6-9(a) has a horizontal force of 80 lb applied to it. Determine if the block will slide up or down the plane or if it will remain stationary. The coefficient of static friction is 0.3.

Since there are three possible actions that the block in the problem could have, each of them must be considered separately. The free-body diagram is shown in Fig. 6-9(b) for the block moving up the plane.

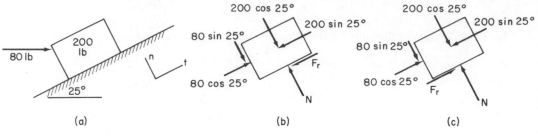

Figure 6-9

If the force up the plane is larger than the maximum friction force plus the component of weight down the plane, the block will slide up the plane. If the force is smaller, the block will remain either stationary or slide down the plane. The normal force is found first by summing forces in the n direction.

$$\Sigma F_n = 0$$

$$N - 80 \sin 25° - 200 \cos 25° = 0$$

$$N = 33.8 + 181.2$$
$$= 215 \text{ lb}$$

Assuming motion up the plane

$$\Sigma F_t = 0$$

$$80 \cos 25° - 200 \sin 25° - F_r = 0$$

$$F_r = 72.5 - 84.5$$
$$= -12.0 \text{ lb}$$

Since $F_r = -12.0$ lb, it means that to just maintain equilibrium the friction force must push down the plane.

The free-body diagram for the block at rest or about to slide down the plane is shown in Fig. 6-9(c).

Assuming the block either slides down the plane or remains at rest

$$\Sigma F_t = 0$$

$$80 \cos 25° - 200 \sin 25° + F_r = 0$$

$$F_r = 84.5 - 72.5$$
$$= 12.0 \text{ lb}$$

If motion impends

$$F_r = \mu N$$
$$= 0.3 \times 215$$
$$= 64.5 \text{ lb}$$

Since the actual friction force of 12 lb is smaller than the friction force of 64.5 lb which must be overcome for motion to start, the block will remain at rest on the plane.

For this problem the graphical solution is probably quicker and simpler than the analytical solution. The solution is shown in Fig. 6-10. First the two known forces of 80 and 200 lb are drawn in their proper directions. Then the normal and friction forces are drawn in their proper directions. Note that there is only one way to draw the friction force so that the force polygon will close. The friction force scaled from the force polygon must be compared with the maximum friction force obtained from $F_r = \mu N$. Since the friction force is smaller than the maximum possible of 64.5 lb, the block will not slide.

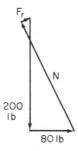

Figure 6-10

PROBLEMS

6-1. If a 40-lb block just starts to slide along a horizontal plane when a horizontal force of 15 lb is applied, what is the coefficient of friction?

6-2. What horizontal force will be required to cause a 50-lb block to just start to move along a horizontal plane if the coefficient of friction is 0.6?

6-3. If the coefficient of friction between a block and a plane is 0.364, what will be the maximum angle to which the plane can be raised before the block starts to slide?

6-4. What is the required coefficient of friction to keep a 200-lb block from sliding down a plane inclined at 35° from the horizontal?

6-5. If a 100-lb weight is on the verge of sliding down a plane inclined at 25° from the horizontal, determine the coefficient of friction.

6-6. If $\mu = 0.2$, calculate the maximum weight which can be supported without slipping of the 40-lb block shown in Fig. P6-6.

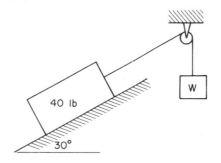

Figure P6-6

6-7. The block shown in Fig. P6-7 weighs 60 lb, and the man weighs 180 lb. If the man is not going to slide, what must be the minimum value for the coefficient of friction between his shoes and the floor?

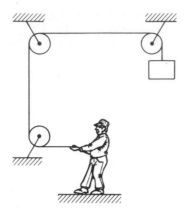

Figure P6-7

6-8. Find the normal force on the block shown in Fig. P6-8.

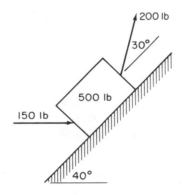

Figure P6-8

6-9. The coefficient of friction between the surface and the 250-lb block shown in Fig. P6-9 is 0.25. Determine the force P which will just initiate motion.

6-10. The block shown in Fig. P6-10 weighs

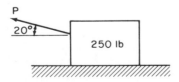

Figure P6-9

50 lb, and the coefficient of friction is 0.30. Determine the force P required to make the block start to move to the left.

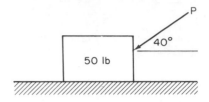

Figure P6-10

6-11. The block shown in Fig. P6-11 weighs 200 lb, the applied force is 100 lb, and μ is 0.30. Show whether or not the block will slide.

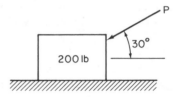

Figure P6-11

6-12. The coefficient of static friction between block A and the plane on which it rests is 0.25. Determine the minimum weight of A for equilibrium for the system shown in Fig. P6-12.

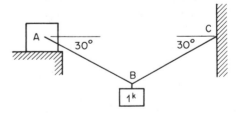

Figure P6-12

6-13. The block shown in Fig. P6-13 weighs 200 lb, and the force P is 40 lb. If the block is about to move down the plane, what is the coefficient of friction?

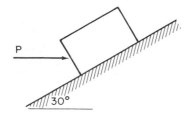

Figure P6-13

6-14. Compute the force P required to cause motion up the plane to just begin in the system shown in Fig. P6-14. There is no friction at pin E, and the pulley is free to rotate.

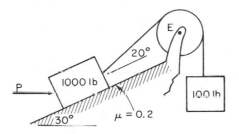

Figure P6-14

6-15. If blocks A and B shown in Fig. P6-15 each weigh W lb, find the minimum value of the coefficient of friction between A and the inclined plane necessary to keep A from sliding.

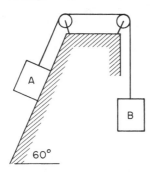

Figure P6-15

6-16. Find the force P required to move the 200-lb block shown in Fig. P6-16. The coefficient of friction for all surfaces is 0.25.

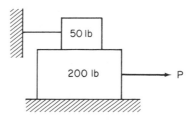

Figure P6-16

6-17. The cylinder shown in Fig. P6-17 weighs 80 lb. The force F is exerted by a rope wrapped around the cylinder. If the cylinder is not to slide down the plane, what must be the coefficient of friction?

Figure P6-17

6-18. Figure P6-18 shows a ladder AB in an alley between two buildings. The ladder weighs 30 lb, and the drum weighs 120 lb. For your convenience the walls of the building are covered with ice ($\mu = 0$). The system is in equilibrium because there is friction at A. Find (a) the reaction of the drum against the vertical wall and (b) the least coefficient of friction possible at A to maintain equilibrium.

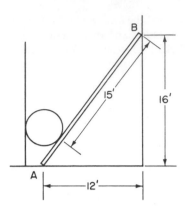

Figure P6-18

6-19. A 2000-lb force is applied to lever AC at C as shown in Fig. P6-19. If the coefficient of friction between surfaces at B is 0.3 determine the largest load W for which equilibrium is possible.

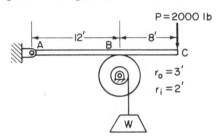

Figure P6-19

6-20. In the brake shoe shown in Fig. P6-20 the coefficient of static friction between the brake shoe and the flywheel is 0.50. Determine the smallest value of P which will prevent rotation of the flywheel if a clockwise couple of 600 ft-lb is applied to the flywheel. Ignore the weight of the brake shoe.

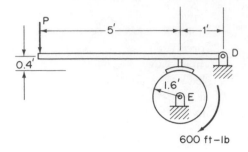

Figure P6-20

6-21. The block E shown in Fig. P6-21 weighs 100 lb. The pin joints are all smooth. (a) Determine the maximum weight W that can be carried without movement of the structure. (b) Calculate the total force at the pin C and the pin D when W is 20 lb.

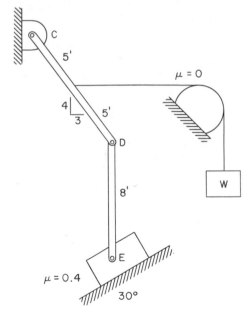

Figure P6-21

6-5. WEDGES

Wedges are one of many mechanical devices that depend on friction for their successful operation. They were probably used by the Egyptians to split stone for use in building the pyramids and are still used today in applications where a small force must be used to produce a small motion and large force. One place where they are still used is in positioning heavy machinery, which must be set as close to horizontal as possible. With a wedge it is possible to carefully control the vertical position much more closely than it can be controlled with any crane. The use of a wedge to lift a box is illustrated in Fig. 6-11. A typical problem would be to find the size of the force P required to raise the box of weight W if the angle of the wedge is θ and the coefficient of friction is μ. The procedure for solving wedge problems is illustrated in Example 6-3. Friction problems are essentially equilibrium problems, and hence free-body diagrams should be used. In fact, for many of the friction problems obtaining the correct solution may be almost impossible without the use of a carefully prepared free-body diagram.

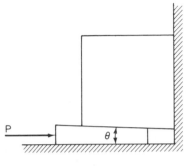

Figure 6-11

EXAMPLE 6-3

A 5° wedge is used to move a 2000-lb weight out from the wall as shown in Fig. 6-12(a). Determine the force P required to move the weight. The coefficient of friction between all surfaces is 0.20.

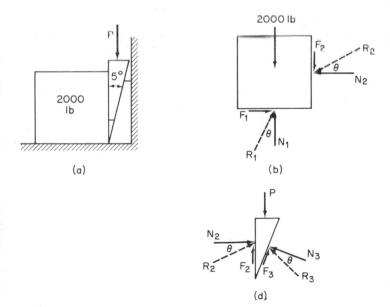

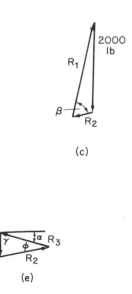

Figure 6-12

To solve this problem a free-body diagram of the weight, as shown in Fig. 6-12(b), should be drawn. Note that since motion is impending on all surfaces, the angle between the normal force and the resultant of the normal and friction forces must be the angle whose tangent is equal to the coefficient of friction.

For impending motion

$$\mu = \tan \theta = 0.20$$
$$\theta = 11.3°$$

Now that we know the orientation of the three forces on the block, the next step is to draw the vector triangle for the forces, as shown in Fig. 6-12(c). In the triangle, R_1 and R_2 are both unknown, but their directions can be easily calculated. The value for R_2 can be found using the sine law.

$$\beta = 90° - \theta - \theta$$
$$= 90° - 11.3° - 11.3°$$
$$= 67.4°$$

$$\frac{R_2}{\sin 11.3°} = \frac{2000}{\sin 67.4°}$$

$$R_2 = \frac{2000 \times \sin 11.3°}{\sin 67.4°} = 425 \text{ lb}$$

If we draw a free-body diagram of the wedge as shown in Fig. 6-12(d), we find that we know R_2 and that we can easily find the directions of the other two forces on the wedge. Again we can draw the vector triangle as shown in Fig. 6-12(e), and we can use the sine law to find the size of P.

$$\alpha = 5° + 11.3° = 16.3°$$
$$\gamma = 90° - 16.3° = 73.7°$$
$$\varphi = 16.3° + 11.3° = 27.6°$$

$$\frac{P}{\sin 27.6°} = \frac{R_2}{\sin 73.7°}$$

$$P = \frac{425 \times \sin 27.6°}{\sin 73.7°} = 205 \text{ lb}$$

Note that the vector triangles in Fig. 6-12 might as well be drawn to scale, so that we obtain the graphical solution at the same time we work out the analytical solution.

6-6. SCREWS

A long continous wedge wrapped around a column is really the basic form of the screw. A screw can be used for lifting an object—as part of the jack of a car—or it can be used to pull parts together—on bolts. The following discussion on forces involved in a screw applies to square-threaded screws. These are the most efficient for transmitting forces and, coincidentally, the easiest to analyze.

A screw jack is shown in Fig. 6-13. The weight W is

carried by a cap, and the screw is turned by means of a force F parallel to the ground and applied perpendicular to the end of a rod of length R. The base supports the screw, and the load is raised relative to the base. Figure 6-14(a) shows a diagram of a thread on the nut that forms the collar on the jack of Fig. 6-13. The thread is shown unwound from the nut. The block shown resting on the thread is the thread of the screw. Since the friction force is independent of area, the screw thread has been shown, for convenience, as a block. The length shown is the average length of one unwound thread, and l is the lead length. P is the effective force on the screw causing it to turn.

By means of the free-body diagram of Fig. 6-14(b), we can show that the torque required to raise the load W is

$$T = FR = \frac{W\bar{D}(\mu\pi\bar{D} + l)}{2(\pi\bar{D} - \mu l)} \qquad (6\text{-}3)$$

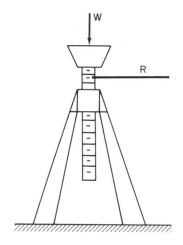

Figure 6-13

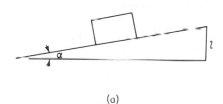

(a)

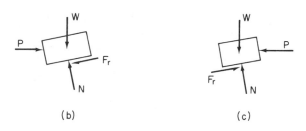

(b)   (c)   **Figure 6-14**

where
    T is the torque required to raise the load,
    F is the force applied to a rod of length R to create the torque,
    W is the load being raised,
    $\bar{D}$ is the average of the inner and outer diameters of the threads,
    μ is the kinetic coefficient of friction, since we are assuming that the load is being raised continuously, and

l is the lead of the screw, the distance a nut would travel along the screw in one complete turn.

If the load is being lowered, the free-body diagram of the screw will be as shown in Fig. 6-14(c). In this case the torque required to lower the load will be

$$T = FR = \frac{W\bar{D}(\mu\pi\bar{D} - l)}{2(\pi\bar{D} + \mu l)} \qquad (6\text{-}4)$$

There is one other case we must consider. It is possible to have a load such that it would lower under its own weight if the coefficient of friction is low enough. For this situation we would still use Eq. (6-4), but the force would be applied so as to prevent lowering of the load. The load would lower under its own weight only if the static coefficient of friction is less than the tangent of the lead angle. The lead angle α is shown in Fig. 6-14(a) and is the angle formed by the lead length and one unwound thread.

EXAMPLE 6-4

A single-threaded screw with a mean diameter of 0.75 in. and a lead of 0.25 in. is used for a jack. If the coefficient of kinetic friction is 0.1, determine what force is required on the end of a 10-in. rod in order to lower a 1200-lb load.

Since the load is to be lowered, we should first check to see that it will not lower by itself.

$$\tan \alpha = \frac{0.25}{0.75\pi} = 0.105$$

We do not know the static coefficient of friction, but because the kinetic coefficient of friction is less than $\tan \alpha$, the load would continue to descend unless restrained by a torque applied to the screw.

From Eq. (6-4) we can find the required force.

$$FR = \frac{W\bar{D}(\mu\pi\bar{D} - l)}{2(\pi\bar{D} + \mu l}$$

$$= \frac{1200 \times 0.75(0.1\pi \times 0.75 - 0.25)}{2(\pi \times 0.75 + 0.1 \times 0.25)}$$

$$F \times 10 = \frac{900(0.236 - 0.250)}{2(2.36 + 0.025)}$$

$$= \frac{450 \times (-.014)}{2.39} = -2.64$$

$$F = -\frac{2.64}{10} = -0.264 \text{ 1b}$$

The force required to lower the load under control or to restrain the load is 0.264 lb.

6-7 BELT FRICTION

Belts are used to transmit a torque. The torque is transmitted by means of the friction force acting between the belt and the pulley.

In Fig. 6-15(a) a belt is shown over a pulley. If it is assumed that T_1 is larger than T_2, then the free-body diagram is as shown in Fig. 6-15(b).

To determine the relationship between the tensions, the coefficient of friction and the angle θ subtended by the belt, calculus is required, so we will not attempt the derivation here. The relationship between the tensions is

$$\frac{T_1}{T_2} = e^{\mu\theta} \qquad (6\text{-}5)$$

where

T_1 is the larger tension,
T_2 is the smaller tension,
e is 2.718, which is called the Naperian constant,
μ is the static coefficient of friction, and
θ is the angle subtended by the belt, measured in radians.

For those who do not have a slide rule with the natural logarithms (a log-log slide rule), Eq. (6-5) may be rewritten as

$$\log \frac{T_1}{T_2} = .434\mu\theta \qquad (6\text{-}5a)$$

This time the logarithms are the common logarithms to the base 10, but the angle θ must still be expressed in radians.

The same principles stated above would apply to band brakes or to determining the holding power of a rope wrapped around a capstan on a quay. When a brake is used to stop a rotating shaft, the kinetic coefficient of friction would be used. In the other cases, as long as there is no slipping, the static coefficient of friction would be used.

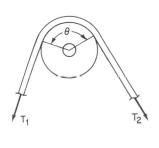

(a)

(b)

Figure 6-15

EXAMPLE 6-5

A rope is wrapped $2\frac{1}{2}$ times around a tree as shown in Fig. 6-16. If the coefficient of friction between the rope and the tree is 0.5, and if the car exerts a force of 600 lb on the rope, determine the pull which must be exerted on the rope to keep the car from moving.

Figure 6-16

This problem is very straightforward; all we have to do is substitute appropriate values in either Eq. (6-5) or (6-5a). For the sake of illustration, we will solve the problem using both equations.

$\theta = 2.5 \times 2\pi = 5\pi$ radians

From Eq. (6-5).

$$\frac{T_1}{T_2} = e^{\mu\theta} = e^{0.5 \times 5\pi}$$

$$T_2 = \frac{600}{e^{7.85}} = \frac{600}{2600} = 0.231 \text{ lb}$$

Using the alternate expression, Eq. (6-5a).

$$\log \frac{T_1}{T_2} = .434\mu\theta$$

$$\log 600 - \log T_2 = .434 \times 0.5 \times 5\pi$$
$$= 3.41$$

$$\log T_2 = 2.778 - 3.41$$
$$= -.632$$
$$= -1 + .368$$

$$T_2 = 0.231 \text{ lb}$$

PROBLEMS

6-22. Both the wedge and lever are used to obtain mechanical advantage. Determine which requires the least force P to lift the weight of the 500-lb box shown in Fig. P6-22. The coefficient of friction is 0.2 for all surfaces. The weight of the box acts through its center.

6-23. Find the weight of the wedge W if

motion is just impending under the weight of the wedge shown in Fig. P6-23.

6-24. Figure P6-24 shows a wedge used to split logs. If the normal force on each face of the wedge is 100 lb, the log will split. If $\mu = 0.3$, what force P must be applied to split the log?

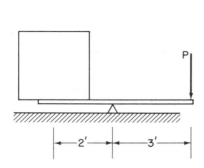

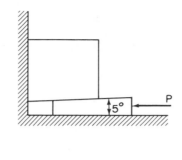

Figure P6-22

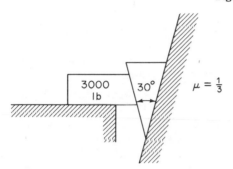

Figure P6-23

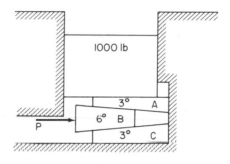

Figure P6-25

1000-lb block and the sidewalls. The coefficient of friction is 0.25 between all other surfaces.

6-26. Find the force P required to remove the wedge shown in Fig. P6-26. The coefficient of friction is 0.25.

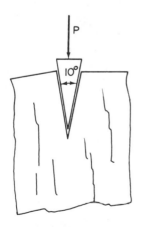

Figure P6-24

6-25. For Fig. P6-25, find the force P exerted on wedge B to raise the 1000-lb block. There is no friction between the

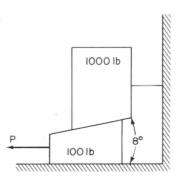

Figure P6-26

6-27. Block *AB* shown in Fig. P6-27 weighs 200 lb and is hinged by a frictionless pin at *B*. The wedge is weightless. Determine the force *P* for impending downward motion of the wedge.

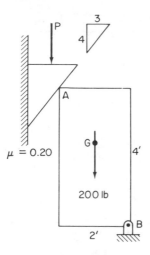

Figure P6-27

6-28. An automobile jack has a screw with an average diameter of 0.825 in. and a lead of 0.30 in. If the coefficient of friction is 0.10, determine the torque that must be applied to the screw to lift an 800-lb load.

6-29. One type of testing machine used to determine the strength of concrete, metals, and other materials applies the load through four large square-threaded screws, each with an average diameter of 3.5 in. and a lead of 0.50 in. If the total load applied to a sample is 40,000 lb, and the coefficient of friction is 0.20, find the torque which must be applied to each screw in order to apply the total load of 40,000 lb.

6-30. For the jack of Prob. 6-28, determine the torque which must be applied to the screw in order to lower the 800-lb load.

6-31. A rod 2 ft long is used to turn a screw with an average diameter of 1.5 in.

and a lead of 0.40 in. If the coefficient of friction is 0.30, and a force of 50 lb is applied at the end of the rod, determine the weight which can be lifted with the screw jack.

6-32. A 2000-lb load is to be lowered using a jack with a screw of 1.0 in. average diameter and a lead of 0.25 in. If the kinetic coefficient of friction is 0.15, determine the force required at the end of a 3-ft rod in order to lower the load.

6-33. What is the maximum pull which can be exerted on one end of a cable which is wrapped two and one half times around a post if the holding force on the other end is 30 lb and the cable is not to slip? The coefficient of friction is 0.20.

6-34. A belt over an 8-in. diameter pulley has a tension of 150 lb on the tight side. If the coefficient of friction is 0.20 and the angle of contact between belt and pulley is 150°, determine (a) the tension of the other part of the belt and (b) the torque transmitted from the belt to the shaft on which the pulley is mounted.

6-35. In order to determine the kinetic coefficient of friction between a brake material and a drum material, a setup similar to that shown in Fig. P6-35 is used. If the tension on scale *A* is 55 lb and the tension on scale *B* is 13 lb, determine the coefficient of friction.

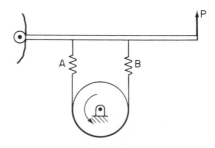

Figure P6-35

6-36. If the coefficient of friction between a rope and a shaft is 0.10, determine the force required on the other end of the rope to hold a force of 2000 lb, if the rope is wrapped 4 times around the shaft.

6-37. Calculate the value of the coefficient of friction required to keep the system shown in Fig. P6-37 in equilibrium. The 5-1b cylinder is smooth.

6-38. Calculate the minimum weight W required to start withdrawal of wedge B shown in Fig. P6-38. The coefficient of friction is 0.25 between all surfaces except at the sidewalls, where it is zero.

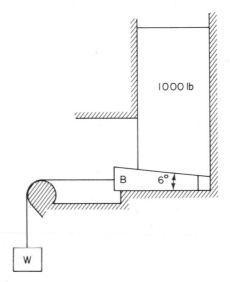

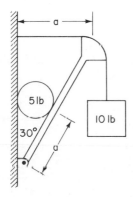

Figure P6-37

Figure P6-38

7

PROPERTIES OF LINES, AREAS, AND SOLIDS

Certain information is required for the solution of some types of applied mechanics problems when the shape of the body influences the way the body behaves.

The information required is the centroid, center of gravity, and moment of inertia. For instance, the amount of bending in a beam depends on the shape of the cross section, as expressed by its area moment of inertia measured with respect to an axis through the centroid, which is the geometric center of the cross section. Some typical beam sections are shown in Fig. 7-1. Also, the tendency of a body to resist rotation depends on its mass moment of inertia, which depends on the mass of the body and the way it is shaped.

7-1. CENTER OF GRAVITY OF PARTICLES IN A PLANE

The location of the center of gravity for any moving body is important, for its location will influence the motion of the body. The distance of the center of gravity above the

155

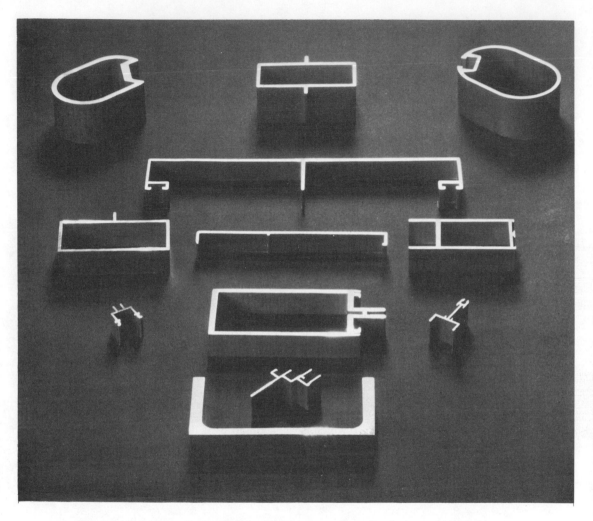

Fig. 7-1. A number of extruded aluminium cross-sections for beams and columns are shown. Before each shape can be used in a structure, the centroid and moment of inertia of the cross-section must be calculated. (Photo courtesy of Alcan Aluminium Limited.)

ground will determine, in part, whether or not a car will overturn on a curve. The location of the center of gravity with respect to the wings and the engines of an airplane will be an important factor in determining whether or not a craft can fly.

Perhaps the easiest way to approach the topic of the center of gravity is to determine the location of the center of gravity of a system of particles of varying weights as shown in Fig. 7-2(a). The center of gravity is the point at which a single

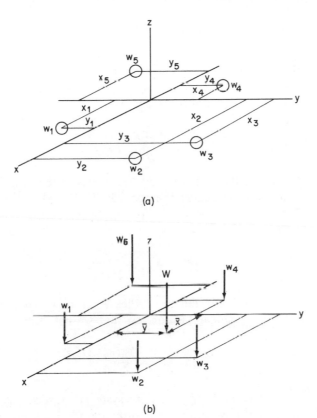

(a)

(b)

Figure 7-2

particle, with a weight equal to the total of the weights of the particles, should be placed so that its attractive force to the earth is the same as the attractive forces of all the particles combined.

A free-body diagram of the particles is shown in Fig. 7-2(b) where the earth's gravitational attraction on each particle is shown as a force. The several forces are to be replaced by a single force located so its effect is the same as the effect of all the other forces. From section 5-4, you will see that this is the same as finding the resultant of a system of parallel forces.

The total weight of all the particles is the resultant force W where

$$W = \Sigma\,(w_i) = w_1 + w_2 + w_3 + \cdots$$

For the effect of the resultant to be the same as that of all the particles, its moment about any axis must be the same as the sum of the moments of the particles about the same axis. Taking moments about the y axis we have

$$
\begin{aligned}
W\bar{x} &= \Sigma\,(w_i x_i) \\
&= w_1 x_1 + w_2 x_2 + w_3 x_3 + \cdots
\end{aligned}
$$

where $\bar{x}$ is the distance from the y axis to the location of the resultant weight, called the x coordinate of the center of gravity. The above expressions are frequently combined in the following form:

$$\bar{x} = \frac{\Sigma\,(w_i x_i)}{\Sigma\,(w_i)} \tag{7-1}$$

The y coordinate of the center of gravity, $\bar{y}$, can be found in a similar fashion, so that we have

$$\bar{y} = \frac{\Sigma\,(w_i y_i)}{\Sigma\,(w_i)} \tag{7-2}$$

The center of gravity of a system of particles is completely defined if we have the coordinates $(\bar{x}, \bar{y}, \bar{z})$. For a system like that shown in Fig. 7-2, where all the particles are in the x-y plane, $\bar{z} = 0$.

In order to determine signs, if the particle is on the positive end of the axis, the product wx or wy may be given a positive sign. If it is at the negative end of the axis, the product wx or wy may be given a negative sign. If we use the above convention, then the sign of the answer that we get for $\bar{x}$ or $\bar{y}$ will be correct, i.e., a positive value tells us that the resultant is placed at the positive end of the axis, and a negative value indicates that the resultant must be at the negative end of the axis.

(The right-hand rule, as used in Chapter 5, could also

be used for our sign convention. However, for center of gravity and centroid problems, the sign convention outlined above is most commonly used.)

EXAMPLE 7-1

Find the center of gravity of the system of particles shown in Fig. 7-3.

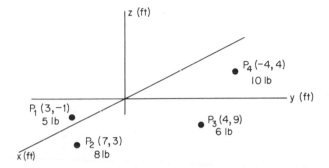

Figure 7-3

The resultant or total weight may be found by summing the weights of the four particles.

$$W = \Sigma (w_i)$$
$$-5 + 8 + 6 + 10$$
$$= 29 \text{ lb}$$

The location of the x coordinate of the center of gravity may be found by solving Eq. 7-1. The sign of each term depends on whether the weight is on the positive or negative end of the axis.

$$\bar{x} = \frac{\Sigma (w_i x_i)}{\Sigma (w_i)}$$
$$= \frac{5 \times 3 + 8 \times 7 + 6 \times 4 - (10 \times 4)}{29}$$
$$= \frac{15 + 56 + 24 - 40}{29}$$
$$= \frac{55}{29}$$
$$= 1.90 \text{ ft}$$

Similarly, $\bar{y}$ may be found by solving Eq. (7-2).

$$\bar{y} = \frac{\Sigma (w_i y_i)}{\Sigma (w_i)}$$
$$= \frac{-(5 \times 1) + 8 \times 3 + 6 \times 9 + 10 \times 4}{29}$$
$$= \frac{-5 + 24 + 54 + 40}{29}$$
$$= \frac{113}{29}$$
$$= 3.90 \text{ ft}$$

7-2. CENTER OF GRAVITY FOR PARTICLES IN SPACE

For a system of particles such as that shown in Fig. 7-4, the center of gravity can be found as easily as finding the center of gravity for particles in a plane. For example, to find

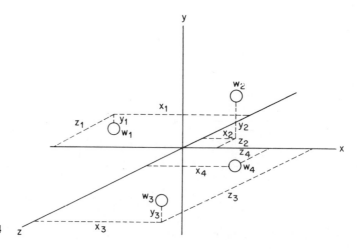

Figure 7-4

$\bar{x}$, moments are taken about the z axis. Note that the distance above the x-z plane does not enter into the moment equation, for the value for $\bar{x}$ does not depend on how far the particles are above the x-z plane. Thus the location of the center of gravity of a system of particles will be given by

$$\bar{x} = \frac{\Sigma \, (w_i x_i)}{\Sigma \, (w_i)} \tag{7-1}$$

$$\bar{y} = \frac{\Sigma \, (w_i y_i)}{\Sigma \, (w_i)} \tag{7-2}$$

$$\bar{z} = \frac{\Sigma \, (w_i z_i)}{\Sigma \, (w_i)} \tag{7-3}$$

where

$\bar{x}, \bar{y}$, and $\bar{z}$ are the coordinates of the center of gravity,
w_i is the weight of the individual particles, and
x_i, y_i, and z_i are the distances to the y-z, x-z, and x-y planes, respectively, for each of the particles.

The sign convention is the same as for particles in a plane.

EXAMPLE 7-2

Find the center of gravity for the system of particles shown in Fig. 7-5.

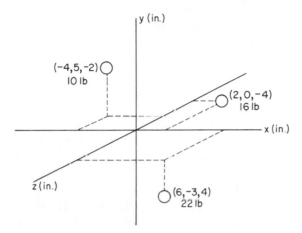

Figure 7-5

The center of gravity is obtained simply by applying Eq. (7-1), (7-2), and (7-3). Care must be taken to see that proper signs are used. The sign will depend on whether the weight is on the positive or negative end of the axis.

$$\bar{x} = \frac{\sum (w_i x_i)}{\sum (w_i)}$$

$$= \frac{10 \times (-4) + 16 \times 2 + 22 \times 6}{10 + 16 + 22}$$

$$= \frac{-40 + 32 + 132}{48}$$

$$= \frac{124}{48} = 2.58 \text{ in}$$

$$\bar{y} = \frac{\sum (w_i y_i)}{\sum (w_i)}$$

$$= \frac{10 \times 5 + 16 \times 0 + 22 \times (-3)}{48}$$

$$= \frac{50 + 0 - 66}{48}$$

$$= \frac{-16}{48} = -0.333 \text{ in.}$$

$$\bar{z} = \frac{\sum (w_i z_i)}{\sum (w_i)}$$

$$= \frac{10 \times (-2) + 16 \times (-4) + 22 \times 4}{48}$$

$$= \frac{-20 - 64 + 88}{48}$$

$$= \frac{4}{48} = 0.0833 \text{ in.}$$

7-3. CENTER OF GRAVITY OF COMPOSITE BODIES

A composite body is a body made up of two or more simple shapes such as cylinders, rectangular parallelepipeds, cones, or spheres. An example of such a body is shown in Fig. 7-6. The centers of gravity of these simple shapes, obtained by precise mathematical calculations, are tabulated in handbooks or in tables such as Table 7-1 at the end of this chapter. The weight of each simple shape is treated as if it acts at the center of gravity of the shape. It is then possible to determine the center of gravity of the whole body by using the weights and centers of gravity of each simple part of the body.

The procedure of dividing a body into several simpler bodies can also be used to obtain an approximate value for the center of gravity of any body. The right circular cone shown in Fig. 7-7(a) can be broken up into many thin cylindrical disks

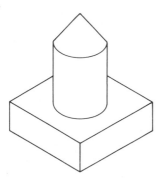

Figure 7-6

 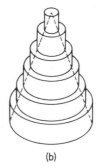

Figure 7-7   (a)   (b)

as shown in Fig. 7-7(b), and the center of gravity can be obtained for the pile of cylindrical disks. The answer will approximate that value tabulated in Table 7-1. If enough very thin disks are used to make up the cone, then the approximate answer and the exact answer will be almost the same.

EXAMPLE 7-3

Find the center of gravity of the bracket shown in Fig. 7-8(a). The material weighs 0.3 lb/in.³

The bracket is divided up into simple shapes whose centers of gravity are already known, as shown in Fig. 7-8(b).

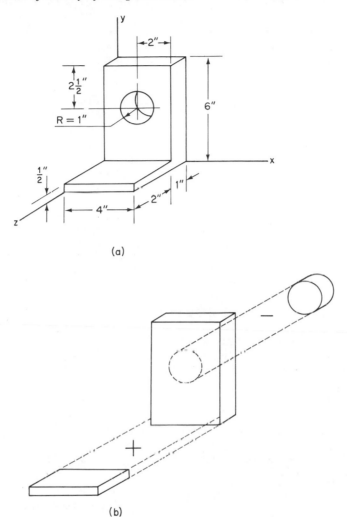

(a)

(b)

Figure 7-8

Although it is not absolutely necessary, it may be most convenient to calculate the weights of each of the parts separately.

Weight of base

$$w_1 = 0.3 \times 0.5 \times 2 \times 4 = 1.20 \text{ lb}$$

Weight of upright

$$w_2 = 0.3 \times 1 \times 4 \times 6 = 7.20 \text{ lb}$$

Weight of hole

$$w_3 = -0.3 \times 1 \times \pi \times 1^2 = -0.945 \text{ lb}$$

The center of gravity may be located by the use of Eqs. (7-1), (7-2), and (7-3).

Because of symmetry

$$\bar{x} = 2.00 \text{ in.}$$

Notice that if a volume is being removed or if distances are on the negative side of the axis, they will be negative terms.

$$\bar{y} = \frac{\sum (w_i y_i)}{\sum (w_i)}$$

$$= \frac{1.20 \times 0.25 + 7.20 \times 3 + (-0.945) \times 3.5}{1.20 + 7.20 - 0.945}$$

$$= \frac{0.30 + 21.60 - 3.31}{7.46}$$

$$= \frac{18.59}{7.46} = 2.59 \text{ in.}$$

$$\bar{z} = \frac{\sum (w_i z_i)}{\sum (w_i)}$$

$$= \frac{1.20 \times 2 + 7.20 \times 0.5 + (-0.945) \times 0.5}{7.46}$$

$$= \frac{2.40 + 3.60 - .473}{7.46}$$

$$= \frac{5.53}{7.46} = 0.741 \text{ in.}$$

PROBLEMS

7-1. Locate the center of gravity of the two particles shown in Fig. P7-1.

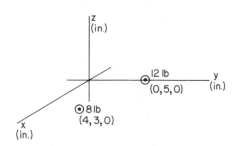

Figure P7-1

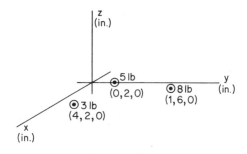

Figure P7-2

7-2. Find the coordinates of the center of gravity of the three particles shown in Fig. P7-2.

7-3. Locate the center of gravity of the system of three weights shown in Fig. P7-3.

7-4. Four particles in the *x-y* plane are shown in Fig. P7-4. Find the center of gravity of the four particles.

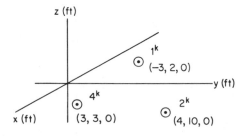

Figure P7-3

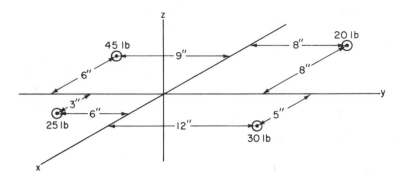

Figure P7-4

7-5. Five particles, A, B, C, D, and G, weigh 25, 35, 40, 15, and 10 lb, respectively, and are all located in the x-y plane at the following locations: $A(2, 3, 0)$, $B(5, -2, 0)$, $C(6, 1, 0)$, $D(0, 0, 0)$, and $G(-3, -4, 0)$, where coordinate distances are in feet. Locate the center of gravity of the system of particles.

7-6. Locate the center of gravity of the system of two particles shown in Fig. P7-6.

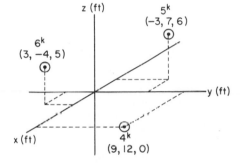

Figure P7-7

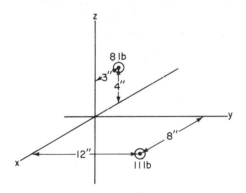

Figure P7-6

7-7. Find the center of gravity for the three particles shown in Fig. P7-7.

7-8. Three particles are shown in Fig. P7-8. Find the center of gravity of the three particles.

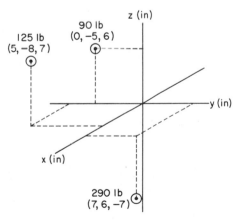

Figure P7-8

7-9. Four particles, A, *B*, *C*, and *D*, weigh 8, 12, 5, and 7 lb, respectively, and are located as follows: $A(5, 3, 9)$, $B(-4, 5, 2)$, $C(6, -4, -3)$, and $D(0, 0, 0)$, where coordinate distances are in inches. Locate the center of gravity of the system of particles.

7-10. The mallet shown in Fig. P7-10 has a rubber head weighing 1.5 lb and a handle weighing 1.0 lb. If the head has a diameter of 3 in. and the handle is 14 in. long, find $\bar{x}$ for the mallet.

7-11. A footing for a large sign is shown in Fig. P7-11. Determine $\bar{z}$ for the footing, which is made of concrete weighing 150 lb/ft³.

7-12. A piston, as shown in Fig. P7-12, is basically a cylinder with a smaller cylinder removed. Calculate $\bar{z}$ for the piston, if the material weighs 0.10 lb/in.³

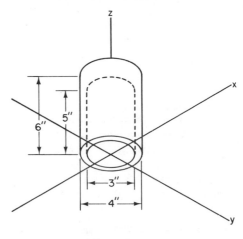

Figure P7-12

7-13. Locate the center of gravity for the bracket shown in Fig. P7-13.

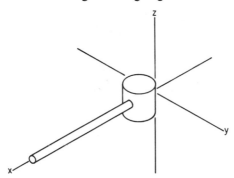

Figure P7-10

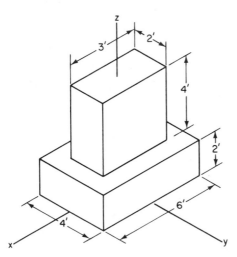

Figure P7-11

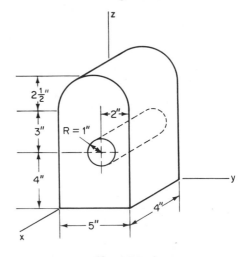

Figure P7-13

7-14. Find the center of gravity of the TV antenna shown in Fig. P7-14. All bars are of the same material and have the same cross-sectional area.

7-15. Three uniform thin plates are joined at right angles to each other as shown in Fig. P7-15. Locate the center of gravity of the composite body.

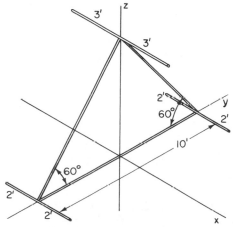

Figure P7-14

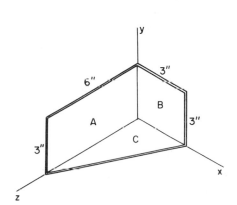

Figure P7-15

7-4. CENTROIDS

The centroid is a geometric property of a line, surface, or volume and represents the central point of the line, area, or body. The location of the centroid for a homogeneous body of one material will, for practical purposes, coincide with the center of gravity.

The approach used to locate the centroid is similar to that used to obtain the center of gravity. For a plane area, the area may be treated as a plate of uniform density ρ and constant thickness t. The surface is divided into many small areas, each of area ΔA, as shown in Fig. 7-9. From Eq. (7-1), the x coordinate of the center of gravity is

$$\bar{x} = \frac{\Sigma\,(w_i x_i)}{\Sigma\,(w_i)} = \frac{\Sigma\,(\Delta A_i\, t \rho x_i)}{\Sigma\,(\Delta A_i\, t \rho)}$$

where $\Delta A_i t \rho = w_i$, the weight of each element. Since t and ρ are constants, they may be taken outside the parentheses and outside the summation sign, so that the common factor can be eliminated from both numerator and denominator. Thus we have

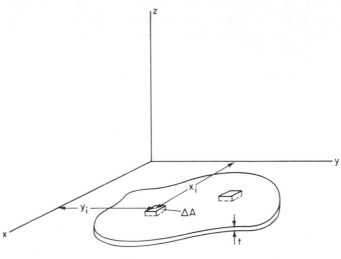

Figure 7-9

$$\bar{x} = \frac{\sum (\Delta A_i \, t\rho x_i)}{\sum (\Delta A_i \, t\rho)} = \frac{t\rho \sum (\Delta A_i \, x_i)}{t\rho \sum (\Delta A_i)}$$

or

$$\bar{x} = \frac{\sum (\Delta A_i \, x_i)}{\sum (\Delta A_i)} \qquad (7\text{-}4)$$

By similar reasoning, we may show that the y coordinate of the centroid is

$$\bar{y} = \frac{\sum (\Delta A_i \, y_i)}{\sum (\Delta A_i)} \qquad (7\text{-}5)$$

In locating centroids it does not matter what shape is chosen for the element of area ΔA as long as the centroid of the element can be readily obtained. For some complex shapes it is best to divide the area into rectangular strips and calculate the location of the centroid. In this case ΔA is the area of each strip, and x (or y) is the distance from the y axis (or x axis) to the centroid of the area of the strip.

In some cases, complex areas may be broken up into a number of simple shapes whose centroids are tabulated in Table 7-1. Here, ΔA is the area of each of the simple areas, and x (or y) is the distance from the y axis (or x axis) to the centroid of the simple area. The distance will have a sign depending on whether the distance to the centroid is a positive or negative distance from the reference axis.

The centroid for a line may be located using a procedure similar to that for locating the centroid of an area. The line is first broken up into simple lines whose centroids are already known. The centroid of the line may then be calculated using the following relationships:

$$\bar{x} = \frac{\Sigma\,(\Delta L_i\,x_i)}{\Sigma\,(\Delta L_i)} \qquad (7\text{-}6)$$

$$\bar{y} = \frac{\Sigma\,(\Delta L_i\,y_i)}{\Sigma\,(\Delta L_i)} \qquad (7\text{-}7)$$

$$\bar{z} = \frac{\Sigma\,(\Delta L_i\,z_i)}{\Sigma\,(\Delta L_i)} \qquad (7\text{-}8)$$

In this case ΔL is the length of the simple line, and x, y, and z are the distances from the reference axes to the centroid of the simple line.

The location of the centroid of a volume can be found using the same procedure as for finding the center of gravity of a body. Instead of using the weights of the parts of the body, however, the volumes of the component parts are used, as indicated in the following equations:

$$\bar{x} = \frac{\Sigma\,(\Delta V_i\,x_i)}{\Sigma\,(\Delta V_i)} \qquad (7\text{-}9)$$

$$\bar{y} = \frac{\Sigma\,(\Delta V_i\,y_i)}{\Sigma\,(\Delta V_i)} \qquad (7\text{-}10)$$

$$\bar{z} = \frac{\Sigma\,(\Delta V_i\,z_i)}{\Sigma\,(\Delta V_i)} \qquad (7\text{-}11)$$

EXAMPLE 7-4

Locate the x coordinate of the centroid for the area shown in Fig. 7-10(a).

The first step is to break the area up into simple shapes whose centroids are known. These simple shapes should approximate the area. The area of Fig. 7-10(a) has been approximated, very roughly, by three rectangles, as shown in Fig. 7-10(b).

Use Eq. (7-4) to determine the location of the centroid of Fig. 7-10(b). Note that

$$\bar{x} = \frac{\Sigma\,(\Delta A_i x_i)}{\Sigma\,(\Delta A_i)}$$

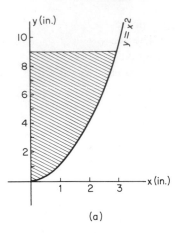

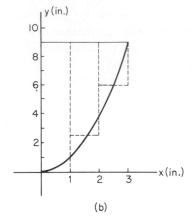

(a)   (b)

Figure 7-10

the values for x are measured from the y axis, not from the base of the rectangle.

$$= \frac{9 \times 1 \times 0.5 + 6.5 \times 1 \times 1.5 + 3 \times 1 \times 2.5}{9 \times 1 + 6.5 \times 1 + 3 \times 1}$$

$$= \frac{4.50 + 9.75 + 7.50}{9.0 + 6.5 + 3.0}$$

If you check the value for $\bar{x}$ from Table 7-1, you will find that the exact answer, obtained using calculus, is 1.125 in. Our approximation is reasonably close.

$$= \frac{21.75}{18.5} = 1.175 \text{ in.}$$

EXAMPLE 7-5

Locate the centroid of the line shown in Fig. 7-11.

First break the line up into simple shapes. The simplest shapes would be (a) the diagonal straight line, (b) the vertical straight line, and (c) the semicircle.

We will simplify the form of our calculations if we calculate the line lengths first.

$$L_a = (6^2 + 3^2)^{1/2} = (36 + 9)^{1/2}$$
$$= (45)^{1/2} = 6.70 \text{ in.}$$
$$L_b = 1.0 \text{ in.}$$
$$L_c = \pi \times 2 = 6.28 \text{ in.}$$

Using the information from Table 7-1, we can use Eqs. (7-6) and (7-7) to locate the centroid of the line.

$$\bar{x} = \frac{\sum (\Delta L_i x_i)}{\sum (\Delta L_i)}$$

$$= \frac{6.70 \times 5 + 1.0 \times 2 + 6.28 \times 4}{6.70 + 1.0 + 6.28}$$

$$= \frac{33.50 + 2.00 + 25.12}{13.98}$$

$$= \frac{60.62}{13.98} = 4.34 \text{ in.}$$

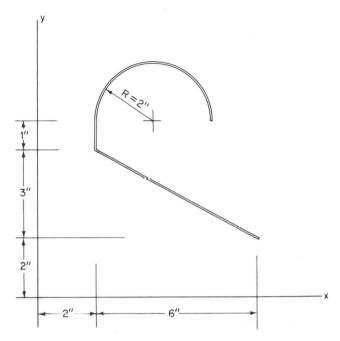

Figure 7-11

The distance from the diameter to the centroid of the semicircle is $2r/\pi$, so that the total distance from the x axis to the centroid is $6 + 2r/\pi$. Note that the centroid of the semicircle can be obtained using two quarter circles.

$$\bar{y} = \frac{\sum(\Delta L_i y_i)}{\sum(\Delta L_i)}$$

$$= \frac{6.70 \times 3.5 + 1.0 \times 5.5 + 6.28 \times (6+1.27)}{13.98}$$

$$= \frac{23.4 + 5.5 + 45.7}{13.98}$$

$$= \frac{74.6}{13.98} = 5.34 \text{ in.}$$

PROBLEMS

7-16. Determine $\bar{y}$ for the built-up T-section shown in Fig. P7-16. (see figure.)

7-17. Find $\bar{x}$ for the angle section shown in Fig. P7-17. (see figure.)

7-18. Find the centroid for the area shown in Fig. P7-18. (see figure.)

7-19. Locate $\bar{x}$ for the area shown in Fig. P7-19. (see figure.)

7-20. Find $\bar{x}$ for the area shown in Fig. P7-20. (see figure.)

7-21. Find $\bar{x}$ for the area shown in Fig. P7-21. (see figure.)

7-22. Calculate $\bar{y}$ for the area shown in Fig. P7-22.

7-23. Find $\bar{y}$ for the area shown in Fig. P7-23.

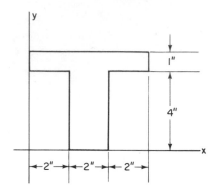

Figure P7-16

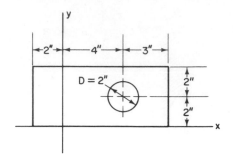

Figure P7-19

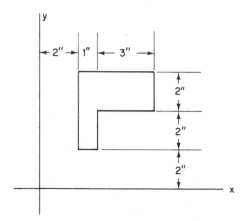

Figure P7-17

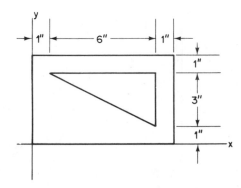

Figure P7-20

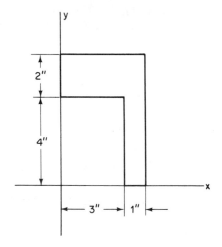

Figure P7-18

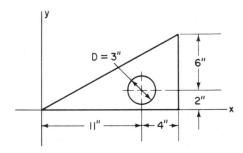

Figure P7-21

7-24. A built-up beam, with the cross-section shown in Fig. P7-24, is made up of an I-beam and a $\frac{3}{4}$-in. plate 6 in. wide welded to the top. If the I-beam is 8 in. deep and has a cross-sectional area of 6.71 in.^{2}, determine the distance from the base to the centroid.

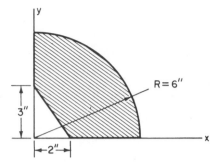

Figure P7-22

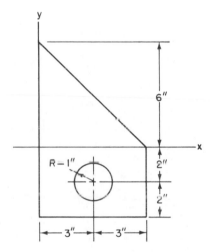

Figure P7-23

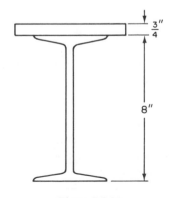

Figure P7-24

7-25. An aluminum extruded mast section is shown in Fig. P7-25. The wall thickness is $\frac{1}{2}$ in. Determine the centroid of the section by (a) treating the section as an area and (b) treating the section as a line along the middle of the wall.

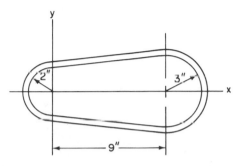

Figure P7-25

7-26. Locate the centroid for the wire shape shown in Fig. P7-26.

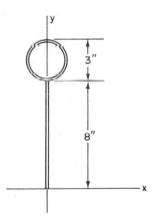

Figure P7-26

7-27. An equilateral triangle, with sides of 8 in. is made of wire. Determine the distance from the base of the triangle to the centroid of the wire.

7-28. Find the centroid of the line shown in Fig. P7-28.

7-29. Locate the centroid of the system of lines shown in Fig. P7-29.

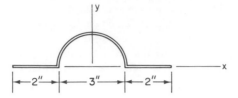

Figure P7-28

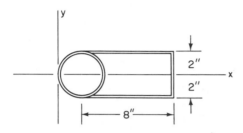

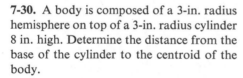

Figure P7-29

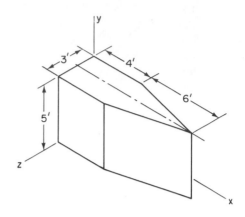

Figure P7-31

7-30. A body is composed of a 3-in. radius hemisphere on top of a 3-in. radius cylinder 8 in. high. Determine the distance from the base of the cylinder to the centroid of the body.

7-31. A storage tank has the shape shown in Fig. P7-31. Determine the centroid of the volume enclosed by the tank.

7-32. Locate the centroid for the shape shown in Fig. P7-32.

7-33. A volume has the shape of a frustrum of a right-circular cone. The base diameter is 8 in., the top diameter is 4 in., and the height of the frustrum is 6 in. Determine the distance to the centroid from the base.

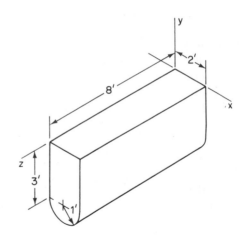

Figure P7-32

7-5. DISTRIBUTED LOADS

Centroids and centers of gravity are used in the process of replacing distributed loads by a single load in the analysis of a structure. Distributed loads might be gravel on the deck of a dump truck, stored materials on the floor of a warehouse, the force of the wind on an airplane's wings, or the load from a floor on a beam.

Distributed loads are frequently represented graphically, as shown in Fig. 7-12 where the distance along the x axis represents distance along the body carrying the load, measured from some reference point. Distances along the vertical axis represent the loading, usually in pounds per foot of length. Sometimes the graph represents loads in pounds per square foot, but we will restrict our discussion to the more common case where the loading is in pounds per foot.

Referring to Fig. 7-12, you will notice that the load distribution starts at 100 lb/ft and eventually increases to about

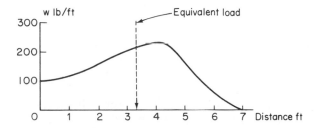

Figure 7-12

230 lb/ft at $x = 4$ ft, and then decreases to zero at the right end. This does not mean that at $x = 4$ ft there is a load of 230 lb over a distance of 1 ft, but if you divided the load in pounds on a very short length by that short length, then the value obtained would be 230 lb/ft. Thus if there is 23 lb on a length of 0.1 ft, then the loading would be 23 lb/0.1 ft $= 230$ lb/ft.

To simplify the analysis of a problem, the distributed load can be replaced by a single load equal to the distributed load and placed at the center of gravity of the distributed load. The total load is equal to the area under the distributed load curve. If you multiply the units along the horizontal axis (feet) by the units along the vertical axis (pounds per foot), you get lb as the resultant units. The center of gravity is at the same point as the centroid of the graph, so that although we talk about the center of gravity of the distributed load, we will in reality be looking for the centroid of the area under the graph. Since we will usually be dealing with a weight, in most cases we will require only the x coordinate of the center of gravity or centroid of the graph.

EXAMPLE 7-6

Replace the distributed load shown in Fig. 7-13 by a single equivalent load.

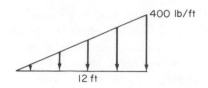

Figure 7-13

The total load is equal to the area under the load diagram, which is triangular.	$W = \frac{1}{2} \times 12 \times 400$ $= 2400 \text{ lb}$
The resultant load is placed at the centroid of the triangle.	$\bar{x} = \frac{2}{3} \times 12 = 8 \text{ ft}$ A single load of 2400 lb is placed 8 ft from the left end.

EXAMPLE 7-7

A beam with a parabolic distributed load is shown in Fig. 7-14(a). Determine the reactions at the two supports.

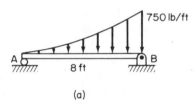

(a)

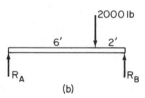

(b)

Figure 7-14

The first step is to determine the total load carried by the beam, which is equal to the area under the parabolic curve.	$W = \frac{1}{3} \times 8 \times 750$ $= 2000 \text{ lb}$
The load is placed at the centroid of the area.	$\bar{x} = \frac{3}{4} \times 8 = 6 \text{ ft}$
To find the reactions, the free-body diagram, Fig. 7-14(b) is used, and moments are taken about any convenient point.	$\sum M_A = 0$ $-(2000 \times 6) + R_B \times 8 = 0$ $R_B = \frac{12000}{8} = 1500 \text{ lb}$
The other reaction is found by summing forces in the y direction.	$\sum F_y = 0$ $R_A - 2000 + 1500 = 0$ $R_A = 500 \text{ lb}$

PROBLEMS

7-34. A uniformly distributed load of 400 lb/ft is placed on a beam 20 ft long. Replace the distributed load by a single concentrated load.

7-35. Determine the single concentrated load which is equivalent to the load shown in Fig. P7-35.

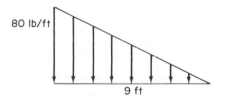

Figure P7-35

7-36. Figure P7-36 shows a distributed load. Replace it by a single equivalent concentrated load.

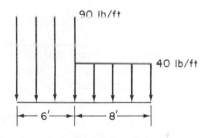

Figure P7-36

7-37. The load on a storage bin floor can be represented as shown in Fig. P7-37. Determine the total weight in the bin and the proper location for the equivalent load.

7-38. Determine the reactions at A and B for a beam with the loading shown in Fig. P7-38.

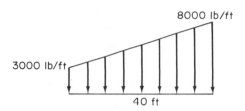

Figure P7-37

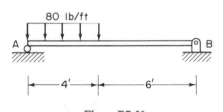

Figure P7-38

7-39. Find the reactions at A and B for the beam shown in Fig. P7-39.

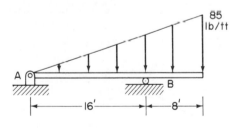

Figure P7-39

7-40. A 1000-ft radio tower is subjected to wind loads as shown in Fig. P7-40. If the tower is not guyed, determine the horizontal force and couple required at the base of the tower to maintain equilibrium.

7-41. An off-the-road dump truck as shown in Fig. P7-41(a) would have its load represented as shown in Fig. P7-41(b). Determine the reaction at support A and pin B.

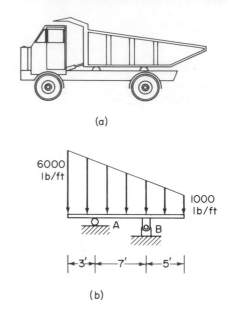

(a)

350 lb/ft

150 lb/ft

Figure P7-40

6000
lb/ft

1000
lb/ft

A

B

|←3'→|← 7' →|← 5' →|

(b)

Figure P7-41

7-6. THEOREMS OF PAPPUS

An interesting application of the use of the centroid occurs in the theorems of Pappus, named after the ancient Greek mathematician who first discovered the relationships. The two theorems permit you to calculate the area of an area of revolution, or the volume of a volume of revolution.

An area of revolution is the area generated by revolving a line, such as that shown in Fig. 7-15, about some axis. The line *ab*, when revolved about the *x* axis, generates the surface shown by the dashed line. The example shown would be like a reducing fitting for a pipeline or the surface of a truncated right-circular cone.

A volume of revolution is shown in Fig. 7-16. The area designated *A* is rotated about the *x* axis. The volume formed is a torus which in this case looks like a rough doughnut.

To find the size of the surface area generated by rotating a line about an axis, consider Fig. 7-17. The surface generated by the very short line segment ΔL about the *x* axis is a hoop which will have a surface area of $2\pi y \, \Delta L$. The total area generated will be the sum of all the hoop areas or $\sum 2\pi y_i \, \Delta L_i$ which may be simplified to $2\pi \sum (\Delta L_i \, y_i)$. From Eq. (7-7) of the

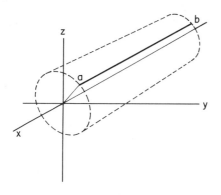

Figure 7-15

discussion on centroids, we have

$$\bar{y} = \frac{\sum (\Delta L_i \, y_i)}{\sum (\Delta L_i)} = \frac{\sum (\Delta L_i \, y_i)}{L}$$

This may be rewritten, by crossmultiplication, as

$$\sum (\Delta L_i \, y_i) = \bar{y}L$$

Thus the surface area generated is

$$A = 2\pi \sum (\Delta L_i \, y_i)$$
$$A = 2\pi \bar{y}L$$

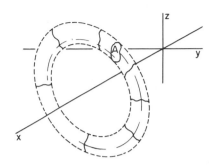

Figure 7-16

where

$\bar{y}$ is the distance from the axis of rotation to the centroid of the line, and

L is the length of the line.

More generally, the area is

$$A = \theta \bar{y}L \qquad\qquad (7\text{-}12)$$

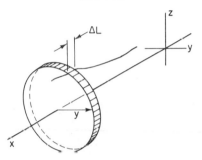

Figure 7-17

where θ is the amount of rotation about the axis expressed in radians, if the amount of rotation is less than 2π radians or $360°$

To find the volume generated by rotating an area about an axis, refer to Fig. 7-18, where the area A, composed of many small areas ΔA, is rotated about the x axis. The volume generated by rotating the one element of area ΔA about the axis is $2\pi y \, \Delta A$. The total volume generated is the sum of the volumes of all the solid hoops, $\sum 2\pi y_i \, \Delta A_i$ or $2\pi \sum (\Delta A_i \, y_i)$. From Eq. (7-5) in our discussion of centroids we have

$$\bar{y} = \frac{\sum (\Delta A_i \, y_i)}{\sum (\Delta A_i)} = \frac{\sum (\Delta A_i \, yi)}{A}$$

By crossmultiplying this may be rewritten as

$$\sum (\Delta A_i \, y_i) = \bar{y}A$$

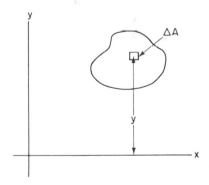

Figure 7-18

Thus the total volume generated is

$$V = 2\pi \sum (\Delta A_i \, y_i)$$
$$V = 2\pi \bar{y}A$$

where

$\bar{y}$ is the distance to the centroid of the generating area measured from the axis of rotation, and

A is the total generating area.

If the area is not rotated through a full 2π radians or 360°, the more general expression for the volume generated is

$$V = \theta \bar{y} A \qquad\qquad (7\text{-}13)$$

where θ is the angle of rotation of the area about the axis, measured in radians.

It should be noted that the theorems of Pappus will not work if either the generating line or the generating area intersect the generating axis.

The theorems of Pappus are perhaps an interesting curiosity. However, they can have practical application for calculating a surface area or a volume of a number of shapes which are not readily available in handbooks.

EXAMPLE 7-8

A cone, as shown in Fig. 7-19(a), is to be made up for a concrete slump test. Determine the amount of material, neglecting overlap, in square inches, required to form the cone and the volume of concrete, in cubic inches, that the cone would hold.

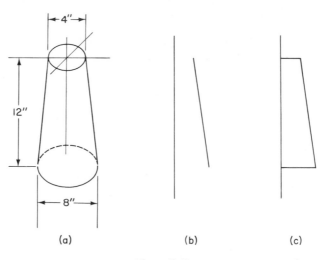

(a)   (b)   (c)

Figure 7-19

To determine the area of the surface we first calculate the length of the line from Fig. 7-19(b) used to generate the slump cone, employing the Pythagorean theorem or the slide rule method for finding the hypotenuse.

$$L = (2^2 + 12^2)^{1/2}$$
$$= (4 + 144)^{1/2} = (148)^{1/2}$$
$$= 12.18 \text{ in.}$$

The centroid of the line will be at its midpoint.

$$\bar{x} = \frac{2 + 4}{2} = 3 \text{ in.}$$

From Eq. (7-12) we can calculate the surface area.

$$A = \theta \bar{x} L$$
$$= 2\pi \times 3 \times 12.18$$
$$= 230 \text{ in.}^2$$

Material required is 230 in.²

To find the volume of material contained by the slump cone, we first calculate the generating area, shown in Fig. 7-19(c).

$$A = \tfrac{1}{2}(2 + 4) \times 12$$
$$= 36 \text{ in.}^2$$

The value for $\bar{x}$ for the generating area can be found using Eq. (7-4) and breaking the area up into a rectangle and a triangle.

$$\bar{x} = \frac{\sum (\Delta A_i x_i)}{\sum (\Delta A_i)}$$
$$= \frac{2 \times 12 \times 1 + \tfrac{1}{2} \times 2 \times 12 \times 2.67}{2 \times 12 + \tfrac{1}{2} \times 2 \times 12}$$
$$= \frac{24 + 32}{24 + 12} = \frac{56}{36}$$
$$= 1.56 \text{ in.}$$

Now Eq. (7-13) may be applied to determine the volume of concrete held by the cone.

$$V = \theta \bar{x} A$$
$$= 2\pi \times 1.56 \times 36$$
$$= 353 \text{ in.}^3$$

Volume of concrete is 353 in.³

PROBLEMS

7-42. How much cardboard, in square inches, is required to make the old-fashioned conical dunce's hat, size 7, 20 in. tall? (Size 7 indicates that the base diameter is 7 in.)

7-43. For the line shown in Fig. P7-43, calculate the surface area generated by rotating the line about the axis.

7-44. Use the theorem of Pappus to cal-

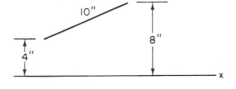

Figure P7-43

culate the volume of the pail and the quantity of material required to make the pail shown in Fig. P7-44.

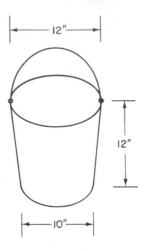

Figure P7-44

7-45. What is the volume of the solid formed by rotating a 5 in. × 12 in. rectangle through 360° about an axis 8 in. from the centroid of the rectangle?

7-46. What volume of rubber is required to produce an O-ring as shown in Fig. P7-46? The cross section of the ring is circular.

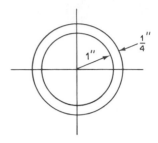

Figure P7-46

7-47. Using the theorem of Pappus, determine the volume of a right-circular cone of height h and radius r. Use a sketch to indicate your procedure.

7-7. AREA MOMENT OF INERTIA

The area moment of inertia, sometimes called the second moment of area, is used in the design and analysis of beams and shafts. The area moment of inertia, frequently called the moment of inertia, is a mathematical expression which occurs in several formula derivations for stress analysis. Because the same mathematical expression occurs frequently, it is subjected to a special study in order to simplify the calculations when it is encountered.

Although the moment of inertia is really a mathematical expression, some people like to have a physical meaning for the term. Two physical expressions for area moment of inertia are (1) a measure of the distribution of area, i.e., whether the area is near to or far from an axis and (2) a measure of the resistance of a beam's cross-section to bending. If you prefer using one of the two tangible meanings, instead of treating the moment of inertia as a mathematical expression, by all means, do so.

Figure 7-20 shows an area, with a small element of area ΔA at coordinates (x, y). For the element shown its moment of inertia about the x axis is

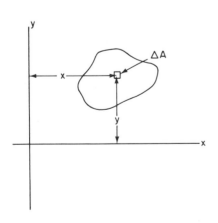

Figure 7-20

$$\Delta I_x = y^2 \, \Delta A$$

This is the mathematical definition of the moment of inertia of the element of area about the x axis. The moment of inertia for the whole area is the sum of the moments of inertia of each of the elements

$$I_x = \Sigma \, (\Delta I_x) = y_1^2 \, \Delta A_1 + y_2^2 \, \Delta A_2 + y_3^2 \, \Delta A_3 + \dots$$
$$I_x = \Sigma \, (y_i^2 \, \Delta A_i) \qquad\qquad (7\text{-}14)$$

Similarly, the moment of inertia of the area about the y axis is

$$I_y = \Sigma \, (x_i^2 \, \Delta A_i) \qquad\qquad (7\text{-}15)$$

Equations (7-14) and (7-15) are the definitions for the moments of inertia of an area about the x and y axes, respectively. As a practical matter, the elements of area chosen are usually narrow strips which *must* be parallel to the axis about which the moment of inertia is being calculated. The distances used are the distances from the axis to the element of area. Although the distance is usually measured to the centroid of the area, this is not strictly correct. Therefore, it is very important that the elements of area be narrow and parallel to the axis with respect to which the moment of inertia is being calculated.

The usual units for area moments of inertia are in.[4], obtained by taking the product of the units of distance squared and area. The moment of inertia will always be positive, because area is positive and the distance, when squared, will always give a positive value. However in cases where we are dealing with a missing area in our calculations, we will treat the moment of inertia of the missing area as if it is negative.

EXAMPLE 7-9

Determine the moment of inertia of the shape shown in Fig. 7-21(a) with respect to the x axis and the y axis.

To find the moment of inertia with respect to the x axis, the area is divided into narrow strips parallel to the x axis as shown in Fig. 7-21(b), and Eq. (7-14) is used. The exact answer, obtained using exceedingly narrow strips, is 97.7 in.[4]

$$I_x = \Sigma \, (y_i^2 \, \Delta A_i)$$
$$= 1.5^2 \times 7 + 2.5^2 \times 7 + 3.5^2 \times 3$$
$$= 15.75 + 43.7 + 36.8$$
$$= 96.3 \text{ in.}^4$$

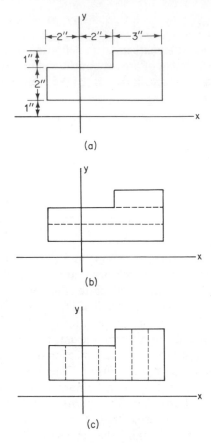

Figure 7-21

The moment of inertia about the y axis is obtained in a similar fashion, except that the elements of area are parallel to the y axis, as shown in Fig. 7-21(c), and Eq. (7-15) is used. The exact answer is 127.6 in.[4]

$$I_y = \sum (x_i^2 \, \Delta A_i)$$
$$= (-1.5)^2 \times 2 + (-.5)^2 \times 2$$
$$+ .5^2 \times 2 + 1.5^2 \times 2$$
$$+ 2.5^2 \times 3 + 3.5^2 \times 3$$
$$+ 4.5^2 \times 3$$
$$= 4.50 + .50 + .50 + 4.50 + 18.75$$
$$+ 36.8 + 60.6$$
$$= 126.2 \text{ in.}^4$$

PROBLEMS

7-48. Find I_x for the area shown in Fig. P7-48.

7-49. Determine the moment of inertia of a rectangle about its base. The rectangle has a height of h and a base of b, and is to be divided into four parallel strips for the calculations.

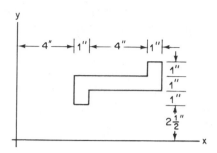

Figure P7-48

7-50. Find I_y for the area shown in Fig. P7-50. Use five strips.

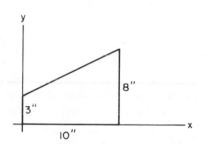

Figure P7-50

7-51. Find the moment of inertia of a circular area about a diameter. Let the circle have a radius of 6 in. and divide it into six strips. A scale drawing would help in determining the lengths of the strips to use in the calculations.

7-52. Figure P7-52 shows a built up I-section for use as a beam. Calculate the moment of inertia of the area about the x axis.

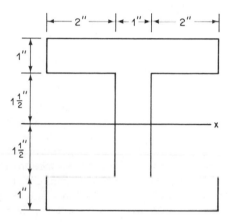

Figure P7-52

7-8. RADIUS OF GYRATION

In column design, an expression called the radius of gyration is frequently used. It is defined as

$$k_x = \sqrt{\frac{I_x}{A}} \qquad (7\text{-}16)$$

where

$\quad$ k_x is the radius of gyration with respect to the x axis,
$\quad$ I_x is the moment of inertia with respect to the x axis, and
$\quad$ A is the area.

If you check the units you will find that the units of radius of gyration are inches if the moment of inertia is in. in.^{4} and the area is in square inches.

The relationship between the moment of inertia and the radius of gyration is more generally written as

$$I_x = k_x^2 A \qquad (7\text{-}17)$$

It should be noted that in many civil engineering handbooks and textbooks, the symbol used for the radius of gyration is *r*.

EXAMPLE 7-10

The moment of inertia of a 35 in.² area about the *y* axis is 184 in.⁴ Determine the radius of gyration of the area about the *y* axis.

The problem is solved simply by making numerical substitutions in Eq. (7-16).

$$k_y = \sqrt{\frac{I_y}{A}}$$
$$= \sqrt{\frac{184}{35}} = \sqrt{5.26}$$
$$= 2.29 \text{ in.}$$

PROBLEMS

7-53. Determine the radius of gyration with respect to the *y* axis for the cross section of a wide flange structural steel section, if the cross-sectional area is 27.63 in.² and the moment of inertia with respect to the *y* axis is 102 in.⁴

7-54. An area of 80 in.² has a moment of inertia about the *x* axis of 600 in.⁴ Calculate the radius of gyration with respect to the *x* axis.

7-55. An area of 16 in.² has a radius of gyration about the *y* axis of 1.2 in. Determine the moment of inertia of the area about the *y* axis.

7-56. Determine the moment of inertia about the *x* axis of an area of 150 in.² if the radius of gyration about the *x* axis is 7.2 in.

7-57. The strength of a column depends in part on its slenderness ratio which is *l/r* where *l* is the length in inches of the column and *r* is the smallest radius of gyration of the cross section. If the smallest moment of inertia of the cross-section of a 20-ft column is 96 in.⁴, determine the slenderness ratio for the column if the area of the cross section is 6 in.²

7-9. PARALLEL AXIS THEOREM

In many cases, the value for the moment of inertia of a particular area is given in tables such as Table 7-1. However, in other cases, the axis that the moment of inertia is taken about for the table is not the same as the one desired. For example, the moment of inertia is frequently given for an axis passing through the centroid of the area, when the desired moment of inertia is about a parallel axis some distance away. In order to clarify some of the figures and discussion which follows, some symbols must be understood.

The symbols $\bar{x}$ and $\bar{y}$, when used as labels for an axis, indicate x or y axes which pass through the centroid. Similarly, $\bar{I}_x$ or $\bar{I}_y$ represent the moments of inertia about the $\bar{x}$ or $\bar{y}$ axes.

The parallel axis theorem, used for finding the moment of inertia about an axis parallel to a given axis, can be derived in the following fashion. Figure 7-22 shows an area A. There are two axes shown, a centroidal axis $\bar{x}$ and a second axis x. The moment of inertia about the x axis is, according to Eq. (7-14),

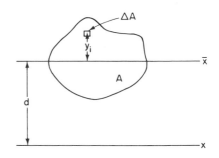

Figure 7-22

$$I_x = \Sigma\,(d + y_i)^2(\Delta A_i)$$
$$= \Sigma\,(d^2 + 2dy_i + y_i^2)\,(\Delta A_i)$$
$$= \Sigma\,(d^2\,\Delta A_i) + \Sigma\,(2dy_i\,\Delta A_i) + \Sigma\,(y_i^2\,\Delta A_i)$$

Since y is measured from the $\bar{x}$ axis, then

$$\Sigma\,(y_i^2\,\Delta A_i) = \bar{I}_x$$

Because the axes are parallel, then d is a constant which can be taken outside of the Σ sign.
Thus

$$\Sigma\,(d^2\,\Delta A_i) = d^2\,\Sigma\,(\Delta A_i) - d^2 A = Ad^2$$

Similarly,

$$\Sigma\,(2dy_i\,\Delta A_i) = 2d\,\Sigma\,(y_i\,\Delta A_i) = 2d\,\Sigma\,(\Delta A_i\,y_i)$$

But from Eq. (7-5)

$$\bar{y}\,\Sigma\,(\Delta A_i) = \Sigma\,(\Delta A_i\,y_i)$$

However, y in Fig. 7-22 is measured from the $\bar{x}$ axis, so that the y distance from the $\bar{x}$ axis to the centroid must be zero, i.e., $\bar{y} = 0$. If $\bar{y} = 0$, then $\Sigma(\Delta A_i\,y_i) = 0$, and thus $2d\,\Sigma(\Delta A_i y_i) = 0$. Thus the moment of inertia of the area of Fig. 7-22 about the x axis is

$$I_x = \bar{I}_x + Ad^2 \qquad\qquad (7\text{-}18)$$

For some reason, as yet unknown, students frequently rewrite Eq. (7-18) to suit their own convenience (incorrectly). It can be rewritten to suit your convenience providing the usual rules of mathematics are followed. For example, if $\bar{I}_x$ is unknown, then it may be found by rewriting Eq. (7-18) as follows:

$$\bar{I}_x = I_x - Ad^2$$

EXAMPLE 7-11

The area shown in Fig. 7-23 is 40 in.^{2}, and has a moment of inertia of 1300 in.^{4} with respect to y_1. Determine the moment of inertia about y_2.

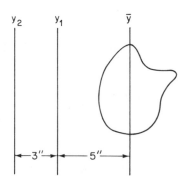

Figure 7-23

In order to use Eq. (7-18), we must first rewrite it to find $\bar{I}_y$.

$$\bar{I}_y = I_{y_1} - Ad^2$$
$$= 1300 - 40 \times 5^2$$
$$= 1300 - 1000$$
$$= 300 \text{ in.}^4$$

Now we may use Eq. (7-18) to find I_{y_2}.

$$I_{y_2} = \bar{I}_y + Ad^2$$
$$= 300 + 40 \times 8^2$$
$$= 300 + 2560$$
$$= 2860 \text{ in.}^4$$

PROBLEMS

7-58. For an area of 55 in.^{2}, $\bar{I}_x$ is 392 in.^{4} Determine I_x if the x axis is 4 in. from the $\bar{x}$ axis.

7-59. Determine the moment of inertia about the $\bar{x}$ axis for an area of 170 in.^{2}, if I_x is 2100 in.^{4} and the two axes are 3 in. apart.

7-60. A centroidal axis for a 12 in.^{2} area is 0.5 in. from the y axis. If it is known that $I_y = 19$ in.^{4}, find I with respect to this centroidal axis.

7-61. A rectangle b units wide by h units high has a moment of inertia of $bh^3/12$ with respect to a centroidal axis parallel to the base. Determine the moment of inertia of the rectangle with respect to its base.

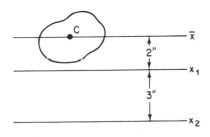

Figure P7-62

7-62. The centroid of the area shown in Fig. P7-62 is at C. If the area is 20 in.^{2} and $I_{x_1} = 110$ in.^{4}, find I_{x_2}.

7-63. An area of 15 in.^{2} has $I_x = 400$ in.^{4} and $\bar{I}_x = 100$ in.^{4} How far apart are the two axes?

7-64. Given that for the area shown in Fig. P7-64, $I_x = 500$ in.^{4} and $\bar{I}_x = 100$ in.^{4}, find the area.

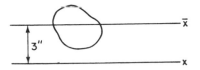

Figure P7-64

7-10. MOMENT OF INERTIA OF COMPOSITE AREAS

In many instances, the moment of inertia is required for an area which can be broken up into several simple shapes whose moments of inertia are known or tabulated in tables such as Table 7-1. In such cases the moment of inertia of each of the parts about the specified axis is summed to find the total moment of inertia. The axis with respect to which the moment of inertia must be calculated is frequently the axis passing through the centroid of the total area and usually differs from the centroids of the various parts of the areas.

The following example problem illustrates the procedure for finding the moment of inertia of a composite area. Note particularly the subtraction of the moment of inertia for an area being removed from one of the areas.

EXAMPLE 7-12

For the area shown in Fig. 7-24(a), determine the moment of inertia about the x axis.

The area is first broken up into simple shapes as shown in Fig. 7-24(b).

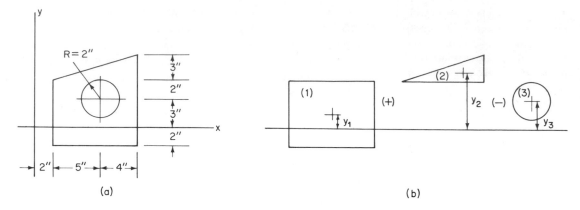

Figure 7-24

The moment of inertia of each of the areas about the x axis is then summed. This will require the use of the parallel axis theorem.

$$I_x = (I_x)_1 + (I_x)_2 - (I_x)_3$$

$$= \frac{9 \times 7^3}{12} + 9 \times 7 \times 1.5^2$$

$$+ \frac{9 \times 3^3}{36} + \frac{1}{2} \times 9 \times 3 \times \left(5 + \frac{3}{3}\right)^2$$

$$- \left(\frac{\pi \times 2^4}{4} + \pi \times 2^2 \times 3^2\right)$$

$$= 257 + 142 + 6.75 + 486$$
$$- (12.6 + 113.3)$$
$$= 891.75 - 125.9$$
$$= 766 \text{ in.}^4$$

There is sometimes a preference for solving problems such as this by using a tabular form as shown below.

Area	$\bar{I}_x$	Ad^2
1	$\dfrac{9 \times 7^3}{12} = \quad 257$	$9 \times 7 \times 1.5^2 = \quad 142$
2	$\dfrac{9 \times 3^3}{36} = \quad 6.75$	$\frac{1}{2} \times 9 \times 3 \times (5 + \frac{3}{3})^2 = \quad 486$
3	$-\dfrac{\pi \times 2^4}{4} = -12.6$	$-\pi \times 2^2 \times 3^2 = -113.3$
Totals	251	515

$$I_x(\text{total}) = \bar{I}_x + Ad^2 = 251 + 515 = 766 \text{ in.}^4$$

PROBLEMS

7-65. Find I_x for the area shown in Fig. P7-65.

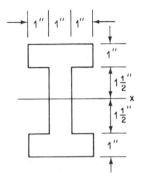

Figure P7-65

7-66. Calculate I_y for the figure shown in Fig. P7-66.

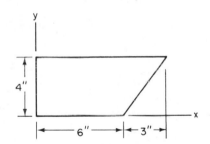

Figure P7-66

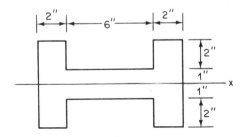

Figure P7-67

7-67. Determine I_x for the shape shown in Fig. P7-67.

7-68. Determine I_x and $\bar{I}_x$ for the area shown in Fig. P7-68.

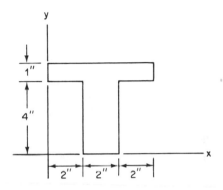

Figure P7-68

7-69. For the section shown in Fig. P7-69 find $\bar{I}_x$

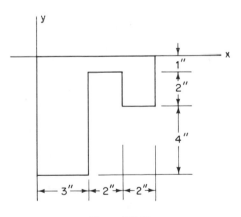

Figure P7-69

7-70. Find $\bar{I}_x$ for the figure shown in Fig. P7-70.

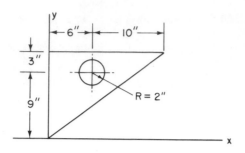

Figure P7-70

7-71. Find $\bar{I}_y$ for the area shown in Fig. P7-71.

7-72. Find I_x for the area shown in Fig. P7-72.

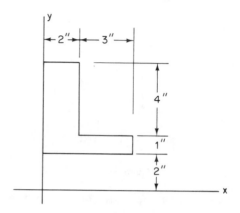

Figure P7-72

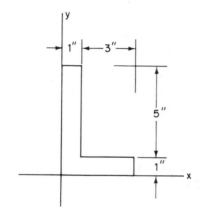

Figure P7-71

7-73. Find $\bar{I}_x$ for the area shown in Fig. P7-73.

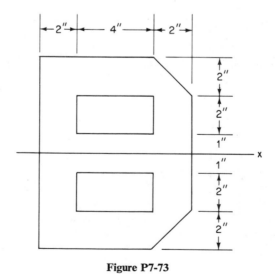

Figure P7-73

7-74. For the area shown in Fig. P7-74(a), $\bar{x} = 1$ in., $\bar{y} = 3$ in., area $= 8$ in.^{2}, $I_x = 87$ in.^{4}, and $I_y = 10$ in.^{4} The section shown in Fig. P7-74(b) is made up of three such areas welded together. Find the centroid and $\bar{I}_x$ for this figure.

(a)

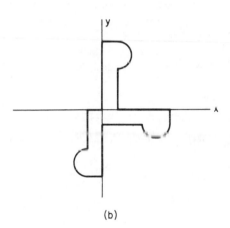

(b)

Figure P7-74

7-75. Find the radius of gyration with respect to the y axis for the area shown in Fig. P7-75.

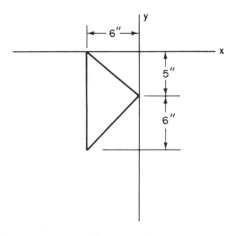

Figure P7-75

7-76. Calculate the radius of gyration of the area shown in Fig. P7-76 with respect to the axis A-A.

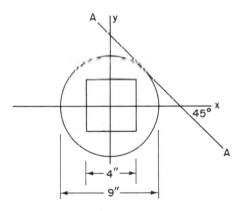

Figure P7-76

7-11. POLAR MOMENT OF INERTIA

A moment of inertia about an axis perpendicular to the plane of the area may be obtained for any area. This is called the polar moment of inertia. It is used in stress analysis for shafts and beams subjected to a torque, for the resistance

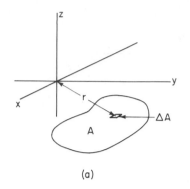

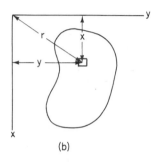

(a)

(b)

Figure 7-25

to twisting depends on the polar moment of inertia. An area in the x-y plane is shown in Fig. 7-25(a). The polar moment of inertia of the element of area ΔA with respect to the z axis is

$$\Delta J_z = r^2 \, \Delta A$$

where

ΔJ_z is the polar moment of inertia of the element of area, and
r is the distance from the area to the z axis.

The total moment of inertia of all the elements of area with respect to the z axis is

$$J_z = \Sigma \, (\Delta J_z) = r_1^2 \, \Delta A_1 + r_2^2 \, \Delta A_2 + r_3^2 \, \Delta A_3 + \cdots$$
$$J_z = \Sigma \, (r_i^2 \, \Delta A_i) \tag{7-19}$$

With the exception of circular shapes which can be broken up into concentric rings, the computations for the polar moment of inertia would be quite time consuming. However, if we refer to Fig. 7-25(b), we observe that

$$r^2 = x^2 + y^2$$

Thus we have

$$J_z = (x_1^2 + y_1^2)\Delta A_1 + (x_2^2 + y_2^2)\Delta A_2 + (x_3^2 + y_3^2)\Delta A_3 + \cdots$$
$$= x_1^2\Delta A_1 + y_1^2\Delta A_1 + x_2^2\Delta A_2 + y_2^2\Delta A_2 + x_3^2\Delta A_3$$
$$\quad + y_3^2\Delta A_3 + \cdots$$
$$= \Sigma \, (x_i^2\Delta A_i) + \Sigma \, (y_i^2\Delta A_i)$$
$$= I_y + I_x$$
$$J_z = I_x + I_y \tag{7-20}$$

In this case, I_x and I_y are the moments of inertia about the x and y axes passing through the perpendicular z axis.

There is a parallel axis theorem for polar moments of inertia. It is

$$J_z = \bar{J}_z + Ad^2 \tag{7-21}$$

where

J_z is the polar moment of inertia with respect to the z axis,
$\bar{J}_z$ is the polar moment of inertia with respect to the $\bar{z}$ axis passing through the centroid of the area,
A is the area, and
d is the distance between the two z axes.

EXAMPLE 7-13

Find the polar moment of inertia for a circle of 3 in. radius for an axis through the centroid.

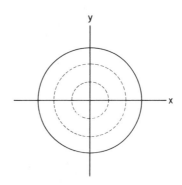

The circle is divided into concentric circles. Three was the number chosen for the solution, as shown in Fig. 7-26.

Figure 7-26

Equation (7-19) may be used to find the polar moment of inertia.

$$\bar{J}_z = \sum (r_i^2 \, \Delta A_i)$$
$$= 0.5^2 \times \pi \times 1^2 + 1.5^2 \times \pi \times (2^2 - 1^2) + 2.5^2 \times \pi \times (3^2 - 2^2)$$
$$= 0.788 + 21.3 + 98.2$$
$$= 120.3 \text{ in.}^4$$

The solution may be obtained exactly by using Eq. (7-20) and the values in Table 7-1 for I_x and I_y.

$$\bar{J}_z = \bar{I}_x + \bar{I}_y$$
$$= \frac{\pi \times 3^4}{4} + \frac{\pi \times 3^4}{4}$$
$$= 63.6 + 63.6$$
$$= 127.2 \text{ in.}^4$$

PROBLEMS

7-77. Find $\bar{J}_z$ for a ring with an inside diameter of 8 in. and an outside diameter of 10 in.

7-78. Find the polar moment of inertia with respect to a centroidal axis of a rectangle 6 in. × 8 in. by dividing the rectangle into four equal rectangles, formed by bisecting the original rectangle vertically and horizontally, and using the defining equation for polar moment of inertia. Check the answer by using the relationship between $\bar{I}_x$, $\bar{I}_y$, and $\bar{J}_z$.

7-79. Determine the polar moment of inertia of the two concentric circular areas, shown in Fig. P7-79, with respect to their centroid.

7-80. For an area in the x-y plane, $I_x = 22$ in.^{4}, and $I_y = 40$ in.^{4} Find the polar moment of inertia.

7-81. An ellipse has semimajor axis a and semiminor axis b. Use the relationship between $\bar{I}_x$, $\bar{I}_y$, and $\bar{J}_z$ to determine the value for $\bar{J}_z$.

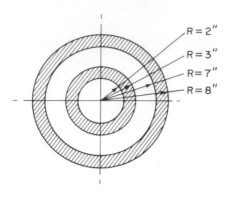

Figure P7-79

7-82. Find the polar moment of inertia of a circle of radius r about an axis through the circumference of the circle.

7-83. The area shown in Fig. P7-83 has the following properties: $A = 4$ in.^{2}, $k_x = 6$ in., and $k_y = 5$ in. Find the polar moment of inertia with respect to an axis through O.

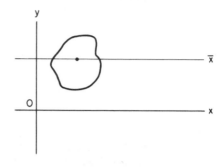

Figure P7-83

7-12. MASS MOMENT OF INERTIA

The moment of inertia of a mass or body does have a physical meaning. It is a measure of the resistance to rotational acceleration of a body and is used in the study of rotating bodies in dynamics.

The mass moment of inertia may be defined mathematically in the following manner. Figure 7-27 shows a body made up of many elements of mass. The moment of inertia of one of these elements of mass, Δm, about the z axis is

$$\Delta I_z = r^2 \, \Delta m$$

where r is the perpendicular distance from the axis to the element of mass.

The total moment of inertia for all the elements of mass in the body would be

$$I_z = \sum \Delta I_z = r_1^2 \, \Delta m_1 + r_2^2 \, \Delta m_2 + r_3^2 \, \Delta m_3 + \cdots$$
$$I_z = \sum (r_i^2 \, \Delta m_i) \qquad (7\text{-}22)$$

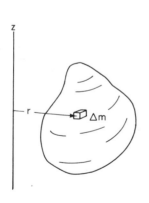

Figure 7-27

The units of mass moment of inertia are usually slug-ft^2. Since the mass is measured in slugs, then the compatible units of length are feet.

Since the units of slugs are lb-sec^2/ft, the units of mass moment of inertia may also be expressed as ft-lb-sec^2.

A slug, in case a bit of review is needed, is a unit of mass. It may be obtained by dividing the weight of the body by the acceleration due to gravity. The gravitational acceleration, g, is about 32.2 ft/sec² on the earth's surface.

A body may be divided into many small bodies, and the mass moment of inertia calculated by the use of Eq. (7-22). However, since these are three-dimensional problems, to obtain sufficient accuracy requires a very large number of calculations. Consequently, we will look primarily at those bodies which can be broken up into simple shapes whose mass moments of inertia have already been precisely calculated.

7-13. MASS MOMENT OF INERTIA OF COMPOSITE BODIES

Bodies made up of several simple shapes, such as those shown in Table 7-1, can be treated in a manner similar to area moments of inertia, where the moments of inertia of the parts about a particular axis are summed to find the total moment of inertia of the entire body about the same axis.

The parallel axis theorem for mass moments of inertia is similar to that for areas.

$$I_x = \bar{I}_x + md^2 \qquad (7\text{-}23)$$

where
I_x is the moment of inertia with respect to the x axis,
$\bar{I}_x$ is the moment of inertia with respect to the $\bar{x}$ axis passing through the center of gravity,
m is the mass of the body and
d is the distance between the two axes.

Caution must be used with the units. Since I will usually be in slug-ft² then m must be in slugs, and d must be in ft.

The radius of gyration is also defined in a manner similar to that used in dealing with areas. For a mass, the radius of gyration is

$$k_x = \sqrt{\frac{I_x}{m}} \qquad (7\text{-}24)$$

This is frequently rewritten as

$$I_x = k_x^2 m \qquad (7\text{-}25)$$

Table 7-1. Properties of Lines, Areas, and Bodies

Shape or Body	Centroid or Center of Gravity of Area or Volume	Moment of Inertia
 Rectangular Parallelepiped	$\bar{x} = 0$ $\bar{y} = 0$ $\bar{z} = 0$ $V = abc$	$I_x = \frac{1}{12}m(a^2 + c^2)$ $I_y = \frac{1}{12}m(b^2 + c^2)$ $I_z = \frac{1}{12}m(a^2 + b^2)$
 Circular Cylinder	$\bar{x} = 0$ $\bar{y} = 0$ $\bar{z} = 0$ $V = \pi r^2 l$	$I_x = \frac{1}{4}mr^2 + \frac{1}{12}ml^2$ $I_y = \frac{1}{2}mr^2$ $I_z = \frac{1}{4}mr^2 + \frac{1}{12}ml^2$
 Slender Rod	$\bar{x} = 0$ $\bar{y} = 0$ $\bar{z} = 0$	$I_x = \frac{1}{12}ml^2$ $I_y = 0$ $I_z = \frac{1}{12}ml^2$
 Right-circular Cone	$\bar{x} = 0$ $\bar{y} = \frac{h}{4}$ $\bar{z} = 0$ $V = \frac{1}{3}\pi r^2 h$	$I_x = \frac{3}{20}mr^2 + \frac{1}{10}mh^2$ $I_y = \frac{3}{10}mr^2$ $I_z = \frac{3}{20}mr^2 + \frac{1}{10}mh^2$

Table 7-1. (Continued)

Shape or Body	Centroid or Center of Gravity of Area or Volume	Moment of Inertia
Sphere	$\bar{x} = 0$ $\bar{y} = 0$ $\bar{z} = 0$ $V = \dfrac{4}{3}\pi r^3$	$I_x = \dfrac{2}{5}mr^2$ $I_y = \dfrac{2}{5}mr^2$ $I_z = \dfrac{2}{5}mr^2$
Hemisphere	$\bar{x} = 0$ $\bar{y} = \dfrac{3}{8}r$ $\bar{z} = 0$ $V = \dfrac{2}{3}\pi r^3$	$I_x = \dfrac{2}{5}mr^2$ $I_y = \dfrac{2}{5}mr^2$ $I_z = \dfrac{2}{5}mr^2$
Rectangle	$\bar{x} = 0$ $\bar{v} = 0$ $A = bh$	$I_x = \dfrac{bh^3}{12}$ $I_v = \dfrac{b^3h}{12}$ $J_z = \dfrac{bh}{12}(b^2 + h^2)$
Triangle	$\bar{x} = 0$ $\bar{y} = 0$ $A = \dfrac{1}{2}bh$	$I_x = \dfrac{bh^3}{36}$
Circle	$\bar{x} = 0$ $\bar{y} = 0$ $A = \pi r^2$	$I_x = \dfrac{\pi r^4}{4}$ $I_y = \dfrac{\pi r^4}{4}$ $J_z = \dfrac{\pi r^4}{2}$

Table 7-1. (Continued)

Shape or Body	Centroid or Center of Gravity of Area or Volume	Moment of Inertia
 Quarter Circle	$\bar{x} = \dfrac{4r}{3\pi}$ $\bar{y} = \dfrac{4r}{3\pi}$ $A = \dfrac{\pi r^2}{4}$	$I_x = \dfrac{\pi r^4}{16}$ $I_y = \dfrac{\pi r^4}{16}$
 Ellipse	$\bar{x} = 0$ $\bar{y} = 0$ $A = \pi ab$	$I_x = \dfrac{1}{4}\pi ab^3$ $I_y = \dfrac{1}{4}\pi a^3 b$ $J_z = \dfrac{1}{4}\pi ab(a^2 + b^2)$
 $y = kx^2$ Parabolic Spandrel	$\bar{x} = \dfrac{3}{4}b$ $\bar{y} = \dfrac{3}{10}h$ $A = \dfrac{bh}{3}$	$I_x = \dfrac{bh^3}{21}$ $I_y = \dfrac{b^3 h}{5}$
 Quartercircular Arc	$\bar{x} = \dfrac{2r}{\pi}$ $\bar{y} = \dfrac{2r}{\pi}$ $L = \dfrac{\pi r}{2}$	

where k_x is the radius of gyration with respect to the x axis.

In order to be compatible with the units of I_x and m, the units of the radius of gyration will usually be feet.

EXAMPLE 7-14

Determine the moment of inertia of the body shown in Fig. 7-28 with respect to the x and z axes. Also calculate k_z. The material in the body weighs 0.3 lb/in.³

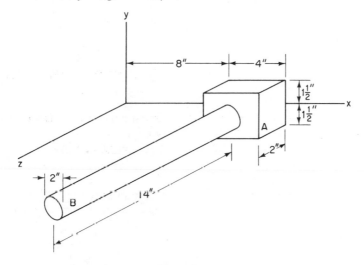

Figure 7-28

The body can be divided into two simple shapes, a rectangular parallelepiped (part *A*) and a cylinder (part *B*).

Although it is not necessary, it is convenient to calculate the mass of each part separate from the moment of inertia calculations.

$$m_A = \frac{2 \times 3 \times 4 \times 0.3}{32.2}$$

$$= 0.224 \text{ slugs}$$

$$m_B = \frac{\pi \times 1^2 \times 14 \times 0.3}{32.2}$$

$$= 0.409 \text{ slugs}$$

The moment of inertia with respect to the x axis of each part can be calculated. Great care must be taken to be certain that correct units are used. Distances must be in feet. Since the x axis does not pass through the centroid of either body, the parallel

With respect to the x axis:
The moment of inertia of A is

$$I_x = \frac{1}{12} \times 0.224\left[\left(\frac{2}{12}\right)^2 + \left(\frac{3}{12}\right)^2\right]$$

$$+ 0.224\left(\frac{1}{12}\right)^2$$

axis theorem must be used. Notice when calculating the moment of inertia of the cylinder, the first term is small enough that we could have neglected the term and used the moment of inertia of a thin rod.

$$= \frac{1}{12} \times 0.224[0.0278 + 0.0625]$$

$$+ 0.001556$$

$$= \frac{1}{12} \times 0.224 \times 0.0903 + 0.001556$$

$$= 0.001686 + 0.001556$$

$$= 0.00324 \text{ slug-ft}^2$$

The moment of inertia of B is

$$I_x = \frac{1}{4} \times 0.409\left(\frac{1}{12}\right)^2 + \frac{1}{12} \times 0.409\left(\frac{14}{12}\right)^2$$

$$+ 0.409\left(\frac{9}{12}\right)^2$$

$$= 0.000709 + 0.464 + 0.230$$

$$= 0.694 \text{ slug-ft}^2$$

Total moment of inertia with respect to the x axis is

$$I_x = 0.00324 + 0.694$$

$$= 0.697 \text{ slug-ft}^2$$

With respect to the z axis:
The moment of inertia of A is

$$I_z = \frac{1}{12} \times 0.224\left[\left(\frac{4}{12}\right)^2 + \left(\frac{3}{12}\right)^2\right]$$

$$+ 0.224\left(\frac{10}{12}\right)^2$$

$$= \frac{1}{12} \times 0.224[0.111 + 0.0625] + 0.155$$

$$= \frac{1}{12} \times 0.224 \times 0.1735 + 0.155$$

$$= 0.00324 + 0.155$$

$$= 0.158 \text{ slug-ft}^2$$

The moment of inertia of B is

$$I_z = \frac{1}{2} \times 0.409\left(\frac{1}{12}\right)^2 + 0.409\left(\frac{10}{12}\right)^2$$

$$= 0.00142 + 0.284$$

$$= 0.285 \text{ slug-ft}^2$$

Total moment of inertia with respect to the z axis is

$$I_z = 0.158 + 0.285$$

$$= 0.443 \text{ slug-ft}^2$$

Equation (7-24) may be used to find k_z.

$$k_z = \sqrt{\frac{I_z}{m}} = \sqrt{\frac{0.443}{0.224 + 0.409}}$$

$$= \sqrt{\frac{0.443}{0.633}} = \sqrt{.700}$$

$$= 0.836 \text{ ft}$$

PROBLEMS

7-84. A flywheel on an engine weighs 644 lb and has a radius of gyration about its axle of 1.25 ft. Determine its moment of inertia.

7-85. A body was found to have a moment of inertia of 12 slug-ft² with respect to the x axis. If the body weighs 1000 lb, find k_x.

7-86. Determine k_y for a body with a mass of 4 slugs and a moment of inertia $I_y = 42$ slug-ft².

7-87. A pulley has a radius of gyration of 3 in. with respect to its axle. If the pulley weighs 8 lb, determine the moment of inertia with respect to the axle.

7-88. The plate shown in Fig. P7-88 is 2 in. thick and weighs 96.6 lb. Calculate its mass moment of inertia with respect to the z axis which passes through the edge of the plate.

7-89. The body shown in Fig. P7-89 consists of an 8-in. high cone on a 4 in. dia-

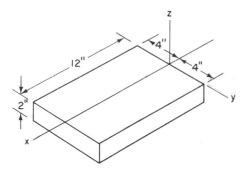

Figure P7-88

meter hemisphere. The material weighs 322 lb/cu ft. Calculate I_r.

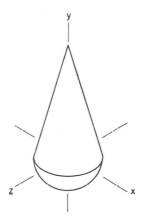

Figure P7-89

7-90. Determine the mass moment of inertia of a hollow cylinder with respect to a longitudinal centroidal axis. The cylinder is 4 ft long, has an outside diameter of 3 ft, and a wall thickness of 4 in. The ends are open. The material weighs 80 lb/ft³.

7-91. The uniform wooden handle of the sledge shown in Fig. P7-91 weighs 3.14 lb and is 3 ft long. The head weighs 16.1 lb. Find the radius of gyration of the sledge with respect to the y axis.

7-92. The wheel shown in Fig. P7-92 has a rim of circular cross section weighing 18 lb and each of the four spokes weigh 7 lb. Determine I_z. [Hint: To find I_z for the rim, use the basic definition for mass moment of inertia, as expressed in Eq. (7-22)].

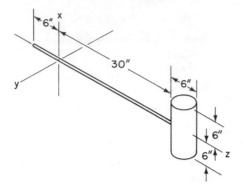

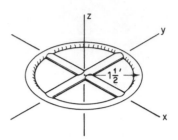

Figure P7-91   **Figure P7-92**

8

KINEMATICS

The study of bodies in motion and the forces that cause motion is called *dynamics*. It has two branches: *kinematics* and *kinetics*. The first branch we will consider is kinematics, the study of the geometry of motion of bodies. Later we will look at kinetics, the study of the forces acting on a body and the motion caused by these forces.

The design of the case unstacker shown in Fig. 8-1 requires the use of the principles of kinematics so that the parts of the machine will not interfere with one another and the cases will be given the desired motion.

8-1. TYPES OF MOTION

Our initial discussion of motion will be restricted to the motion of particles. A *particle* may be defined as any body whose dimensions are small relative to the radius of the path or possible path of the body. Hence a satellite traveling around the earth may be treated as a particle, while a man moving in the satellite would be treated as a body relative to the satellite, not as a particle.

Another way of expressing the difference is to say that if the size of the body affects its motion, then the body must be

205

Fig. 8-1. The design of this case unstacker requires careful analysis of the motion that the cases must have, and the motion that the machine parts must have in order to cause the desired motion of the cases. (Photo courtesy of Standard Conveyor Company.)

treated as a body of finite extent. If the size does not affect the motion, the body may be treated as a particle. For example, a car traveling in a straight line at constant speed may usually be treated as a particle, but a car going around a sharp curve may have to be treated as a body, because the distance of the center of gravity above the road influences how the car will move.

For convenience in discussing motion, we classify motion into several types. The simplest motion is motion in a straight line, as indicated in Fig. 8-2(a). This is also called rectilinear motion. Nearly as simple is circular motion where a particle travels in a circular path about an axis. A particle at point *A* on the pulley shown in Fig. 8-2(b) would have circular motion, as would any other point on the pulley. A third type of motion of a particle is general plane motion where the particle follows some path in a plane which is neither circular nor straight. Such motion is indicated in Fig. 8-2(c). Naturally, motion is not confined to a plane, so there must also be a fourth class of motion which we can call general motion, where a particle travels in space or where all three dimensions must be used to describe its motion.

8-2. DISPLACEMENT

The displacement of a particle is a vector quantity which measures change in position. It is important to emphasize the term change in position; displacement of a particle does not depend on the path taken from the original to the final position. Three different paths. *A,B,* and *C* from P_1 to P_2 are shown in Fig. 8-3. For each of these cases the straight line from P_1 to P_2 is the displacement of the particle when it has moved from P_1 to P_2, even though the paths taken may be considerably different. The displacement is usually represented by **s**. Displacement, like other vectors, can be added and subtracted, as was shown in the initial discussion on vectors in Section 3-3.

Distance is a scalar and is a measure of the length of path traveled. Referring to Fig. 8-3 the magnitude of the displacement from P_1 to P_2 is the length of line *B*, even though the particle traveled along line *C*. In this case the distance traveled by the particle would be much longer than the magnitude of the displacement.

The units of displacement and distance will be length units, usually inches, feet, or miles.

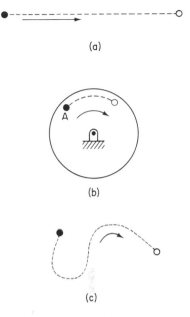

(a)

(b)

(c)

Figure 8-2

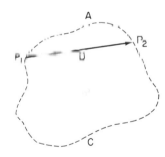

Figure 8-3

8-3. VELOCITY

There are several ways of defining the velocity of a particle. One is that the velocity is the rate of change of displacement with respect to time. A more correct definition is

$$\mathbf{v} = \frac{\Delta \mathbf{s}}{\Delta t} \tag{8-1}$$

which may also be written as

$$\mathbf{v} = \frac{\mathbf{s}_2 - \mathbf{s}_1}{t_2 - t_1}$$

where

$\mathbf{v}$ is the velocity of the particle,

$\Delta \mathbf{s}$ is the change in displacement, which is equal to $\mathbf{s}_2 - \mathbf{s}_1$, and

Δt is the change in time, which is equal to $t_2 - t_1$.

Equation (8-1) is most accurate for very small values of Δt. It is important to realize that an average velocity is obtained when the value of Δt is large relative to the total amount of time, and instantaneous velocity is obtained for values of Δt so small that they approach zero.

From the notation, and from the fact that velocity is a change in displacement divided by a scalar, it should be obvious that when we use velocity, we are dealing with a vector, for it has both magnitude and direction. The direction will be the same as the change in displacement.

Speed is a scalar and is simply the magnitude of the velocity.

The units of velocity will be length per unit of time, usually in./sec, ft/sec, or mi/hr.

EXAMPLE 8-1

At $t = 3$ sec a particle has a displacement of 3 in. from the origin 30° clockwise from the positive x axis, and at $t = 5$ sec it has a displacement of 7 in. upward along the positive y axis. Determine the displacement of the particle during the time interval from $t = 3$ sec to $t = 5$ sec and the average velocity of the particle during the same period.

Figure 8-4

Since we are dealing with vectors, drawing a vector triangle, as shown in Fig. 8-4, would seem to be the first step. The triangle should represent $s_5 - s_3$, since $\Delta s = s_5 - s_3$.

The cosine law is then used to determine the magnitude of Δs.

$$(\Delta s)^2 = (s_3)^2 + (s_5)^2 - 2(s_3)(s_5) \cos 120°$$
$$= 3^2 + 7^2 - 2 \times 3 \times 7 \times (-0.5)$$
$$= 9 + 49 + 21$$
$$\Delta s = 79^{1/2} = 8.90 \text{ in.}$$

To completely define Δs, we need its direction. This can be found by using the vector triangle and the sine law. It is a good idea to make the figures to scale so that the answers may be checked graphically.

$$\frac{\sin \theta}{s_3} = \frac{\sin 120°}{\Delta s}$$
$$\sin \theta = \frac{3 \times \sin 120°}{8.90}$$
$$= 0.292$$
$$\theta = 17.0°$$

The displacement during the period $t = 3$ sec to $t = 5$ sec is

8.90 in.   73.0° ⟍

The average speed can be found using Eq. (8-1) expressed in scalar form. To convert this to velocity, it is necessary to take into consideration the direction of Δs, since the direction of v must be the same as Δs.

$$v = \frac{\Delta s}{\Delta t}$$
$$= \frac{8.90}{2} = 4.45 \text{ in./sec}$$
$$v = 4.45 \text{ in./sec} \quad 73.0° ⟍$$

8-4. ACCELERATION

The acceleration of a particle may be defined as the rate of change of its velocity with respect to time. However, the more correct mathematical defintion is

$$\mathbf{a} = \frac{\Delta \mathbf{v}}{\Delta t} \qquad\qquad (8\text{-}2)$$

which may also be rewritten as

$$\mathbf{a} = \frac{\mathbf{v}_2 - \mathbf{v}_1}{t_2 - t_1}$$

where

$\mathbf{a}$ is the acceleration,

$\Delta\mathbf{v}$ is the change in velocity, equal to $\mathbf{v}_2 - \mathbf{v}_1$, and

Δt is the change in time, which is equal to $t_2 - t_1$.

An average value for acceleration is obtained if Δt is large relative to the total time. If Δt is so small that it approaches zero, then the instantaneous acceleration is obtained. The acceleration obtained above is a vector. Unfortunately, the magnitude of acceleration also is called acceleration, which is a scalar. Usually it is possible to tell from the statement whether the scalar or vector acceleration is being referred to. The units of acceleration are length per unit time per unit time, usually in./sec², ft/sec², or mi/hr².

EXAMPLE 8-2

An airplane has a velocity of 500 ft/sec north 25° east at one instant, and 4 seconds later has a velocity of 500 ft/sec due east. Find its average acceleration during this period of time.

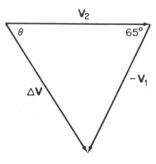

Figure 8-5

According to Eq. (8-2) we need the change in velocity to find the acceleration, so the vector diagram, showing $\Delta\mathbf{v} = \mathbf{v}_2 - \mathbf{v}_1$ is drawn, as shown in Fig. 8-5. Notice that even though the speed does not change, the velocity changes because of the plane's change in direction.

The cosine law may be used to find the magnitude of $\Delta\mathbf{v}$. If our vector triangle was drawn to scale, we have a graphical check on our calculations.

$$(\Delta v)^2 = (v_1)^2 + (v_2)^2 - 2(v_1)(v_2)\cos 65°$$
$$= 500^2 + 500^2$$
$$- 2 \times 500 \times 500 \times \cos 65°$$
$$= 250{,}000 + 250{,}000 - 212{,}000$$
$$= 288{,}000$$
$$\Delta v = 288{,}000^{1/2} = 536 \text{ ft/sec}$$

Now the scalar form of Eq. (8-2) may be used to find the magnitude of **a**.

$$a = \frac{\Delta v}{\Delta t}$$

$$= \frac{536}{4} = 134 \text{ ft/sec}^2$$

We still require the direction to completely define the acceleration. It may be obtained by using the cosine law or by observing that we have an isosceles triangle in this particular problem.

$$\theta = \frac{180° - 65°}{2} = 57.5°$$

$$\mathbf{a} = 134 \text{ ft/sec}^2 \qquad 57.5° \ \diagdown$$

PROBLEMS

8-1. A ship moves 30 mi south 25° west then it moves 18 mi north 75° west. Determine (a) its final displacement from its original position and (b) the distance traveled.

8-2. A point on the circumference of a 2-ft radius pulley moves from a position directly above the axle to a position directly below the axle. Determine the initial and final displacements of the point measured from the center of the pulley and the distance traveled by the point.

8-3. A particle travels from point $A(3, 7)$ to $B(4, 8)$ to $C(-3, 2)$. Determine the displacement from A to B, from B to C, from A to C, and the total distance traveled. Distances are in feet.

8-4. In a period of 3 sec, a body has a displacement of 45 ft south 20° east. Determine the average velocity of the particle during this period.

8-5. At $t = 3$ sec a particle has a displacement of 18 in. from the origin 30° above the positive x axis, and at $t = 4$ sec the particle has a displacement of 12 in. from the origin 60° above the negative x axis. Determine the average velocity of the particle.

8-6. A point on a flywheel rotating in a vertical plane is 3 ft from the center of rotation. If the point is initially on the same level as the center, determine the average velocity of the point if in 0.02 sec it has moved through an arc of 3°. Also solve the problem for when the point has moved through an arc of 90° in 0.6 sec.

8-7. A particle moves to the right in a horizontal straight line. At $t = 7$ sec it passes through point $A(4, 6)$, and at $t = 11$ sec it passes through $B(25, 6)$. Determine the displacement from the origin to B and the average velocity of the particle as it travels from A to B. Distances are in feet.

8-8. A particle has a velocity of 40 ft/sec parallel to the positive x axis and 0.5 sec later has a velocity of 25 ft/sec parallel to the positive y axis. Determine the average acceleration of the particle during this time interval.

8-9. A car is traveling north at 35 mi/hr and in 3 sec accelerates to 60 mi/hr north. Determine the acceleration of the car in (a) ft/sec² and (b) mi/hr².

8-10. A point on the left side of the circumference of a 2-ft diameter pulley has a velocity upward of 35 ft/sec when it is on the same level as the center of the pulley. When the pulley has revolved clockwise through 5°, the velocity is 55 ft/sec directed 5° clockwise from the positive y axis. If the motion takes 0.3 sec, determine the average acceleration during this time interval.

8-11. The velocity of a particle at $t = 4$ sec is 140 ft/sec at 45° counterclockwise from the positive x axis, and at $t = 7$ sec it is 25 ft/sec at 15° counterclockwise from the positive y axis. Calculate the average acceleration during the time interval.

8-12. Figure P8-12 shows three displacement vectors for a particle. If $t_1 = 2$ sec, $t_2 = 5$ sec, and $t_3 = 7$ sec, determine the average velocities for the two time intervals. Use these average velocities to find an average acceleration during the total time interval. (Note that the value obtained for average acceleration will be a rough approximation since average velocities are used in the computation.)

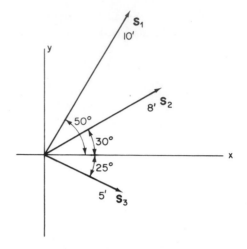

Figure P8-12

8-5. LINEAR MOTION WITH CONSTANT ACCELERATION

One of the more common types of motion, and one of the easiest to analyze, is linear motion with constant acceleration. Bodies in a vertical free fall near the surface of the earth have this kind of motion provided they do not fall long distances.

Relationships between time, distance, speed, and acceleration can be developed using the definitions given by Eqs. (8-1) and (8-2). Since we are dealing with linear motion, the scalar values will be used, and direction will be indicated by the use of the plus or minus sign.

From Eq. (8-2) we have

$$a = \frac{\Delta v}{\Delta t}$$

or

$$a = \frac{v_2 - v_1}{\Delta t}$$

If we crossmultiply, this becomes

$$v_2 - v_1 = a(\Delta t)$$

or

$$v_2 = v_1 + a(\Delta t) \qquad (8\text{-}3)$$

This shows that the final velocity of a particle is equal to the initial velocity plus the product of the acceleration and the elapsed time, provided the acceleration is constant and the particle is traveling in a straight line.

From Eq. (8-1) we have the following value for the average velocity

$$v = \frac{\Delta s}{\Delta t}$$

or

$$v = \frac{s_2 - s_1}{\Delta t}$$

The average velocity is also equal to

$$v_{av} - \frac{v_2 + v_1}{2}$$

But from Eq. (8-3),

$$v_2 = v_1 + a(\Delta t)$$

Substituting for v_2 in our equation for v_{av}

$$v_{av} = \frac{v_1 + a(\Delta t) + v_1}{2} = v_1 + \frac{1}{2}a(\Delta t)$$

From Eq. (8-1) we have the value for average velocity

$$v_{av} = \frac{s_2 - s_1}{\Delta t}$$

By crossmultiplying we have

$$s_2 - s_1 = v_{av}(\Delta t)$$

If we substitute in our value for average velocity we have

$$s_2 - s_1 = [v_1 + \tfrac{1}{2}a(\Delta t)]\Delta t$$
$$s_2 = s_1 + v_1(\Delta t) + \tfrac{1}{2}a(\Delta t)^2 \qquad (8\text{-}4)$$

From Eq. (8-3) we may solve for Δt

$$\Delta t = \frac{v_2 - v_1}{a}$$

This may be substituted in one of the modified forms of Eq. (8-1)

$$s_2 - s_1 = v_{av}(\Delta t)$$

so that we have

$$s_2 - s_1 = \left(\frac{v_2 + v_1}{2}\right)\left(\frac{v_2 - v_1}{a}\right)$$

$$s_2 - s_1 = \frac{v_2^2 - v_1^2}{2a}$$

This may be rewritten as

$$v_2^2 - v_1^2 = 2a(s_2 - s_1) \qquad (8\text{-}5)$$

In using the three equations developed above, emphasis must be placed on the fact that we are always measuring changes in speed, distance, or time, and that values for speed, distance, and time can be positive or negative, depending on where the body is relative to the origin, on whether it is moving in a positive or negative direction, and on the time relative to the starting time. In many problem solutions, such as when the particle starts from rest, or from the origin, one or more of the values in the equations will be zero. It cannot be overemphasized that Eqs. (8-3), (8-4), and (8-5) are valid only if the acceleration is constant. Zero acceleration is just a special case of constant acceleration.

The most common constant acceleration is that produced by gravity. It has a value of 32.2 ft/sec² near the earth's surface and is directed to the center of the earth.

EXAMPLE 8-3

A ball is thrown vertically upward with an initial speed of 40 ft/sec. How high will the ball go and how long will it take to return to the point it was thrown from?

We may determine the height that the ball rises from Eq. (8-5). We assume that the ball starts from the origin, and when it

$$v_2^2 - v_1^2 = 2a(s_2 - s_1)$$
$$0 - (40)^2 = 2 \times (-32.2)(s_2 - 0)$$
$$-1600 = -64.4s_2$$

reaches its maximum height its speed must be zero, for the ball has to stop to change direction and fall.

$$s_2 = \frac{1600}{64.4} = 24.8 \text{ ft}$$

It is possible to determine the time taken for the ball to rise and fall by using Eq. (8-4). Notice that the distance from the starting point at both the beginning and end of travel is zero. The time could also have been obtained using Eq. (8-3), if it is recognized that the speed on return is the same as when leaving the thrower, except for the direction.

The ball rises to a height of 24.8 ft
$$s_2 = s_1 + v_1(\Delta t) + \tfrac{1}{2}a(\Delta t)^2$$
$$0 = 0 + 40(\Delta t) + \tfrac{1}{2}(-32.2)(\Delta t)^2$$
$$0 = 40 - 16.1(\Delta t)$$
$$\Delta t = \frac{40}{16.1} = 2.48 \text{ sec}$$

Total time required is 2.48 sec

If we keep in mind that displacement, velocity, and acceleration are vectors, and that vectors can be added, we will find that there is a group of problems involving two-dimensional motion which can be solved readily using the methods outlined above.

The problems are projectile problems where the motion in two directions can be described in terms of motion with constant acceleration in each direction. The two components of motion at any particular time are added vectorially to determine the total motion of the particle.

EXAMPLE 8-4

Gravel is raised into a bin by means of a conveyor as shown in Fig. 8-6. If the gravel is carried at 15 ft/sec by the conveyor, determine (a) how long it takes the gravel to hit the floor of the bin, (b) the horizontal distance from the end of the conveyor to the floor of the hopper or bin where the gravel strikes, and (c) the velocity at which the gravel strikes the floor.

There is no acceleration in the horizontal direction. But there is a constant acceleration due to the earth's gravity in the vertical direction.

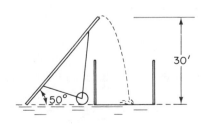

Figure 8-6

If we can find out how long it takes the gravel to hit the floor of the bin, we can determine how far the particle travels horizontally in the same period of time. The origin used is the top of the conveyor.

(a) Initial speed in the y direction

$$v_{y_1} = 15 \cos 40°$$
$$= 11.5 \text{ ft/sec}$$
$$s_2 = s_1 + v_1(\Delta t) + \tfrac{1}{2}a(\Delta t)^2$$
$$-30 = 0 + 11.5(\Delta t) + \tfrac{1}{2}(-32.2)(\Delta t)^2$$
$$-30 = 11.5(\Delta t) - 16.1(\Delta t)^2$$
$$16.1(\Delta t)^2 - 11.5(\Delta t) - 30 = 0$$

Solve using the standard solution form for quadratic equations:

$$\Delta t = \frac{-b \pm \sqrt{b^2 - 4ac}}{2a}$$
$$= \frac{11.5 \pm \sqrt{(11.5)^2 - 4 \times 16.1 \times (-30)}}{2 \times 16.1}$$
$$= \frac{11.5 \pm \sqrt{132.2 + 1932}}{32.2}$$
$$= \frac{11.5 \pm 44.7}{32.2}$$
$$= 1.74 \quad \text{or} \quad -1.03 \text{ sec}$$

Since the starting time was when the gravel left the conveyor, the positive value of the solution is the correct one.

The floor is struck after 1.74 sec

Once the time of travel is known, the distance traveled in the x direction can be calculated using Eq. (8-4).

(b) Speed in the x direction is constant and is

$$v_{x_1} = 15 \sin 40°$$
$$= 9.65 \text{ ft/sec}$$
$$s_2 = s_1 + v_1(\Delta t) + \tfrac{1}{2}a(\Delta t)^2$$
$$= 0 + 9.65 \times 1.74 + 0$$
$$= 16.8 \text{ ft}$$

The gravel strikes the floor 16.8 ft from the end of the conveyor in the horizontal direction.

To find the velocity at which the gravel strikes the floor, we need to find both the x and y components of velocity at $t = 1.74$ sec.

(c) In the vertical direction

$$v_{y_2} = v_{y_1} + a(\Delta t)$$
$$= 11.5 - 32.2 \times 1.74$$
$$= 11.5 - 56.2$$
$$= 44.7 \text{ ft/sec}$$

The speed in the x direction is constant.

$$v_{x_2} = 9.65 \text{ ft/sec}$$

The velocity may be found using the Pythagorean theorem or the slide rule method for finding resultants.

$$v_2 = 45.7 \text{ ft/sec} \quad \searrow 77.8°$$

PROBLEMS

8-13. An automobile accelerates at 15 ft/ sec² for 5 sec. If it is traveling at 25 ft/sec at the beginning of the 5 sec, what is its speed at the end of 5 sec, and how far does it travel in this time?

8-14. A stone is dropped a distance of 100 ft to the ground from the tenth floor of a building. How long will it take the stone to hit the ground? If the stone is dropped from the twentieth floor (200 ft up) how long will it take the stone to hit the ground?

8-15. A car traveling at 60 mi/hr should be able to stop in a distance of 300 ft after the brakes are applied on a dry highway. Assuming that the acceleration (sometimes called *deceleration*) is constant, determine the necessary acceleration in ft/sec².

8-16. A baseball is "popped up" by a batter. If the vertical component of velocity is 60 ft/sec, will a spectator 40 ft above the batter have a chance to catch the ball? If not, how high will the ball go?

8-17. An automobile engine has a maximum rpm of 4200 and a stroke of 4 in. If the acceleration is constant from the middle of the stroke to the end of the stroke, deter-

mine the acceleration in ft/sec² during one-quarter of the piston's cycle.

8-18. A bullet traveling horizontally at 1500 ft/sec leaves the barrel of a rifle 5 ft above the ground. How long will it take the bullet to strike the ground, and how far will it travel?

8-19. A stone is dropped from a building on to a steel drum on the ground. If sound travels at 1100 ft/sec, determine the distance from the point where the stone was dropped to the drum if the sound was heard by the person that dropped the stone 8 sec after the stone was released.

8-20. If a stream of water from a fire hose is to reach the top of a 50-ft building, determine the required vertical component of the velocity of the water when it leaves the nozzle.

8-21. A baseball is struck and leaves the bat at 75 mi/hr and at an angle of 40° above the horizontal. If the ball is struck 5 ft above the ground, will it clear a 12-ft fence 370 ft from the batter's box? If the ball does not hit the fence, where will it land?

8-6. GRAPHICAL SOLUTION OF LINEAR MOTION PROBLEMS

Kinematics problems involving linear or straight-line motion can be solved using graphical or analytical methods. A somewhat greater variety of problems can be solved conveniently using the graphical procedures because it is not necessary to restrict our solutions to problems where the acceleration is constant.

The following procedures will illustrate the relationships which can be obtained graphically among the distance, speed, acceleration, and time.

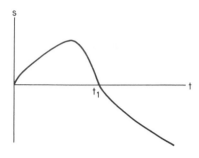

Figure 8-7

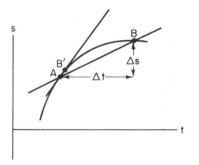

Figure 8-8

The distance a particle moves from some fixed reference point is frequently plotted relative to time on an *s-t* or displacement-time graph similar to the one shown in Fig. 8-7. The graph shown indicates a particle starting at the origin and traveling along the positive axis. At time t_1, it has returned to the origin and then it starts traveling along the negative axis.

For linear motion, the graph shown in Fig. 8-7 can be used to calculate velocity or speed. Speed is the magnitude of velocity: $v = \Delta s/\Delta t$. A small part of an *s-t* graph is shown in Fig. 8-8 along with the values for change in distance Δs and the change in time Δt. The ratio $\Delta s/\Delta t$ is the slope of the line joining points *A* and *B*.

As Δt becomes smaller, the length of Δs will also change, and *B* will move over to *B'*. As Δt gets smaller, line *AB'*, when extended, will become a tangent to the curve when *A* and *B'* are so close together that they coincide.

The average speed of the particle between *A* and *B* is the slope of the line *AB*. The instantaneous speed of the particle, when *A* and *B'* coincide, is the slope of the tangent to the curve at *A*. Thus the speed of the particle at any time is equal to the slope of the *s-t* graph at that time.

Realizing that $v = \Delta s/\Delta t$, it is possible now to construct a velocity-time or *v-t* graph as shown in Fig. 8-9. A series of tangents is drawn at intervals along the *s-t* curve. The slope of each of these tangents is then calculated. This can be done most quickly if the Δt interval chosen is one time unit such as 1 second or 1/10 second, since it is very easy to divide with numbers such as these. The values calculated for *v* can be plotted on a *v-t* graph as shown in Fig. 8-9. It is both practical and customary to plot the two graphs one under the other as shown.

Using reasoning similar to that used above, we can construct an acceleration-time or *a-t* graph from a *v-t* graph. From Fig. 8-10 we see that the average acceleration between points *C* and *D* is $a = \Delta v/\Delta t$ which is the slope of the line connecting *C* and *D*. As Δt becomes very small, and *D* moves to *D'* the line *CD'* will become tangent to the curve at *C*. The slope of the tangent to the curve at *C* is then the instantaneous acceleration of the particle at the time corresponding to point *C*.

The same method that was used for plotting the velocity-time graph can now be used for plotting an acceleration-time graph. The tangents to the *v-t* graph are drawn at convenient intervals, usually seconds, or some fraction of a

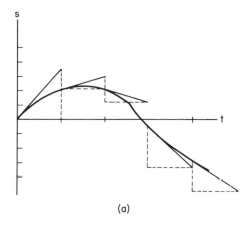

(a)

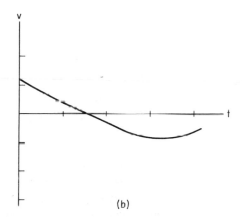

(b)   **Figure 8-9**

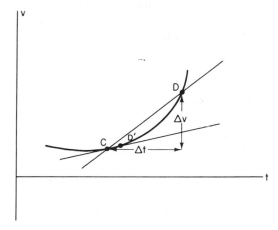

Figure 8-10

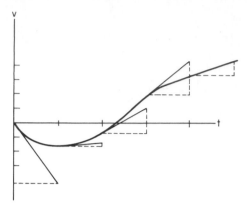

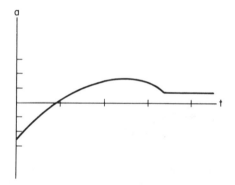

Figure 8-11

second, which is convenient for division. The slopes of the tangents are then calculated and plotted as the acceleration-time curve as shown in Fig. 8-11.

It is both common practice and convenient to plot the *a-t* graph under the *v-t* graph. In fact, if all three graphs are being drawn together, they are usually plotted one under the other.

The are a few points which should be observed with regard to the diagrams shown in Figs. 8-9 and 8-11. Notice that where the upper curve reaches a maximum or minimum value, the lower curve crosses the axis. This should be expected since the slope of a maximum or minimum on a graph is zero. At any region where the upper curve is a straight line, the lower curve will be a horizontal straight line. This is due to the fact that the constant slope of the straight line in the upper curve yields a constant value for the lower curve. Also, where there is an abrupt change in the shape of the upper curve, an extra slope should be obtained at that point in order to improve the accuracy of the lower curve being derived.

EXAMPLE 8-5

A displacement-time graph is shown in Fig. 8-12(a). Use the graph to draw a velocity-time graph and an acceleration-time graph.

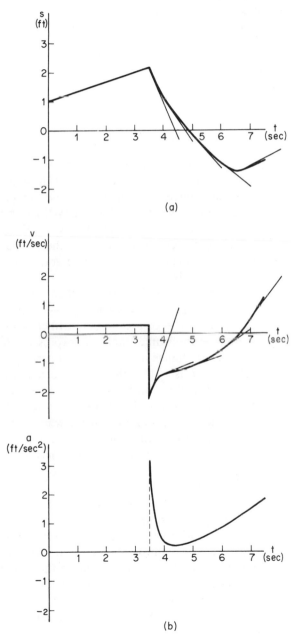

(a)

(b)

Figure 8-12

The axis system for the *v-t* graph is aligned under the *s-t* graph. Tangents are drawn at 1-second intervals and at abrupt changes in the curve, as occurs at 3.5 sec. Since Δt is 1 second for each tangent, the values for v will be equal to the values of Δs. These are plotted for the *v-t* graph. When the *v-t* graph is completed, the same procedure is then used to draw the *a-t* graph below. Notice that at 3.5 sec of the *v-t* graph the slope is infinity. This value is rather unrealistic. The peak value for acceleration is obtained from the slope of the curve just beyond the vertical line.

In the above discussion we see that it is possible, for linear motion, to plot a velocity-time graph if we have the displacement-time graph, and that we can plot the acceleration time graph if we have the velocity-time graph. Now we shall determine how it is possible to reverse the process and plot a velocity-time graph, if we know the acceleration-time graph, and obtain the displacement-time graph, if we have the velocity-time graph.

The expression for acceleration, $a = \Delta v/\Delta t$ may be rewritten as $\Delta v = a(\Delta t)$. The expression $a(\Delta t)$ is approximately equal to the shaded area in Fig. 8-13 between times t_1 and t_2 on the acceleration-time graph. The expression $\Delta v = a(\Delta t)$ then tells us that the change in velocity from time t_1 to t_2 is $a(\Delta t)$. If the velocity at t_1 is v_1, then the velocity at t_2 will be $v_2 = v_1 + \Delta v = v_1 + a(\Delta t)$. This can be plotted as a velocity-time graph from t_1 to t_2 as shown in Fig. 8-13. If the velocity at the origin, or any other point, is known, then a whole series of areas can be calculated by counting squares in a graph or approximating the area by using a trapezoid. Each new area is added to the preceding value for velocity and used to plot the full velocity-time graph.

In a similar fashion it is possible to draw a displacement time graph using the velocity-time graph as shown in Fig. 8-14. The expression $v = \Delta s/\Delta t$ can be rewritten as $\Delta s = v(\Delta t)$, where $v(\Delta t)$ is approximately the area of the shaded portion of the *v-t* curve. This area is equal to the change in displacement from time t_1 to t_2. If some initial displacement is known, then by evaluating a large number of areas representing change in displacement, a complete *s-t* graph can be drawn.

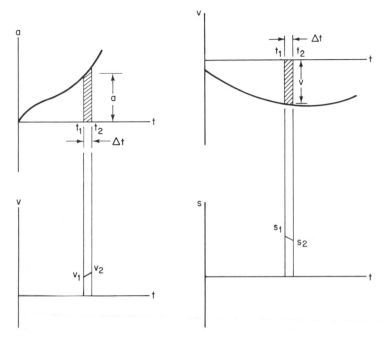

Figure 8-13   **Figure 8-14**

The shape of the curve will govern how large an increment of area or time interval to use for evaluating either Δs or Δv. If the curve is bounded by straight lines, it is usually safe to use the whole area beneath the single straight line for the calculation. If the curve contains many abrupt changes, then the time increments for which areas are calculated must coincide with the many abrupt changes in the curve.

The areas used in forming both *v-t* and *s-t* graphs have signs. If the area is above the *t* axis, it is positive, and if it is below the *t* axis, it is negative. A positive area represents a positive change in velocity or displacement, and a negative area represents a negative change in velocity or displacement.

EXAMPLE 8-6

Using the *a-t* graph shown in Fig. 8-15(a) determine the *v-t* and *s-t* graphs if the speed at $t = 0$ sec is -20 ft/sec, and the displacement at $t = 0$ sec is zero.

The graphical solution is shown in Fig. 8-15(b). The necessary calculations are shown below.

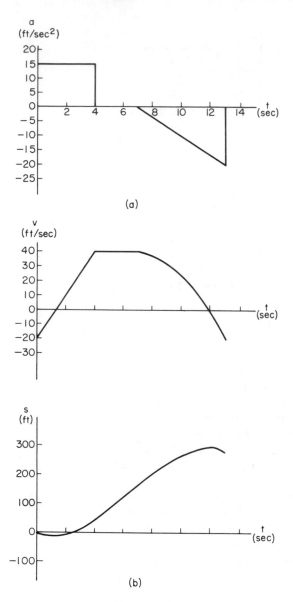

(a)

(b)

Figure 8-15

The change in speed is equal to the area under the *a-t* graph. Notice that since the initial speed is -20 ft/sec, the *v-t* graph must start from -20 ft/sec. Since the acceleration is constant during the first 4 sec, the *v-t* graph must have a constant slope.

To construct the *v-t* graph for $t = 0$ to $t = 4$ sec

$$\Delta v = 15 \times 4 = 60 \text{ ft/sec}$$

Thus at $t = 4$ sec

$$v = -20 + 60 = 40 \text{ ft/sec}$$

There is no area under the *a-t* graph from 4 to 7 sec, so the speed does not change during this time period.

For $t = 4$ to $t = 7$ sec

$\Delta v = 0$

Thus at $t = 7$ sec

$v = 40$ ft/sec

From 7 to 13 sec the area under the *a-t* graph is triangular. This means that the slope of the *v-t* graph will be constantly changing from zero at 7 sec to -20 at 13 sec. The curve formed will be a parabola.

For $t = 7$ to $t = 13$ sec

$\Delta v = -(\frac{1}{2} \times 6 \times 20) = -60$ ft/sec

Thus at $t = 13$ sec

$v = 40 - 60 = -20$ ft/sec

Since it is the area between the graph and the axis that represents the change in distance, we must first determine where the *v-t* graph crosses the *t* axis. This can be done using the similar triangles between 0 and 4 sec of the *v-t* graph.

To construct the *s-t* graph for $t = 0$ to $t = t_1$ sec, by similar triangles:

$$\frac{t_1}{20} = \frac{4 - t_1}{40}$$

$40t_1 = 80 - 20t_1$

$60t_1 = 80$

$t_1 = 1.33$ sec

The *s-t* graph starts from zero and has a change of -13.3 ft. The curve will be parabolic and will reach a maximum where the *v-t* graph crosses the *t* axis.

For $t = 0$ to $t = 1.33$ sec

$\Delta s = -(\frac{1}{2} \times 1.33 \times 20) = -13.3$ ft

Thus at $t = 1.33$ sec

$s = -13.3$ ft

The change from 1.33 to 4 sec continues the parabola, but the values increase positively.

For $t = 1.33$ to $t = 4$ sec

$\Delta s = \frac{1}{2} \times 2.67 \times 40 = 53.3$ ft

Thus at $t = 4$ sec

$s = -13.3 + 53.3 = 40.0$ ft

From 4 to 7 sec, speed is constant, so the distance traveled increases linearly.

For $t = 4$ to $t = 7$ sec

$\Delta s = 40 \times 3 = 120$ ft

Thus at $t = 7$ sec

$s = 40 + 120 = 160$ ft

From 7 to 13 sec, the *v-t* graph is parabolic and crosses the *t* axis. We must first determine when the *v-t* graph crosses the *t* axis. Since $\Delta v = a(\Delta t)$, we can find the base of the triangle in the *a-t* graph which will be

For $t = 7$ to $t = t_2$ sec from the similar triangles of Fig. 8-16

$$\frac{a}{t} = \frac{20}{6}$$

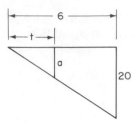

Figure 8-16

just large enough to give a Δv of 40 ft/sec. Similar triangles from the *a-t* graph are used to solve for the value of *t*.

$$a = \frac{20}{6}t$$

For $\Delta v = 40$

$$40 = \frac{1}{2}t \times \frac{20}{6}t$$

$$t^2 = \frac{40 \times 2 \times 6}{20} = 24$$

$$t = 4.90 \text{ sec}$$

Thus $t_2 = 7 + 4.90 = 11.90$ sec

The formula for the area of a parabola must be used to find Δs.

$$\Delta s = \frac{2}{3} \times 4.90 \times 40 = 130.8 \text{ ft}$$

Thus at $t = 11.90$ sec

$$s = 160 + 130.8 = 290.8 \text{ ft}$$

The graph from 11.90 to 13 sec is a continuation of the parabola, but the curve is flat enough so that the area can be approximated by a triangle, saving much computation.

For $t = 11.90$ to $t = 13$ sec

$$\Delta s = -(\tfrac{1}{2} \times 1.10 \times 20) = -11.0 \text{ ft}$$

Thus at $t = 13$ sec

$$s = 290.8 - 11.0 = 279.8 \text{ ft}$$

PROBLEMS

8-22. A particle starts at the origin and has a constantly increasing linear displacement-time curve, so that at 10 sec it has traveled 300 ft. Use graphical procedures to draw the *v-t* graph for the first 10 sec.

8-23. A body starts moving in a straight line with a speed of 40 ft/sec, and in 4 sec its speed is decreased at a constant rate to 10 ft/sec. Use graphical procedures to draw the *a-t* graph.

8-24. The *s-t* graph for a body can be described by the equation $s = -7 + 3t^2$, where *s* is in miles, and *t* is in hours. Use graphical methods to find the acceleration-time graph for the first 5 hrs.

8-25. The *s-t* graph for a particle is semicircular in shape. The time interval for the graph is 24 sec, and the maximum distance, reached at 12 sec, is 200 ft. Use this information to draw the *v-t* and *a-t* graphs.

8-26. Draw the *v-t* graph for a particle using the *s-t* graph shown in Fig. P8-26.

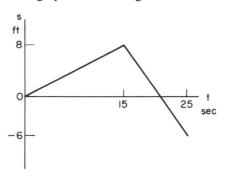

Figure P8-26

8-27. Find the *v-t* and *a-t* graphs for a particle which has the *s-t* graph shown in Fig. P8-27.

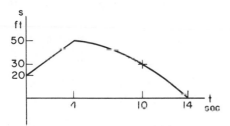

Figure P8-27

8-28. Use the graph shown in Fig. P8-28 to determine the *v-t* and *a-t* graphs for the body whose displacement is described by the graph.

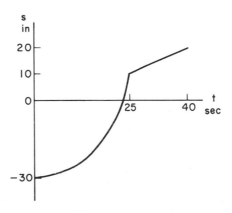

Figure P8-28

8-29. Cams are used in many machines to give a part a required displacement, speed, or acceleration. The cam shown in Fig. P8-29(a) will cause the cam follower to have the displacement shown in Fig. P8-29(b). Use this information to determine the *v-t* and *a-t* graphs for the follower.

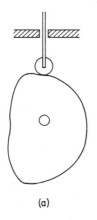

(a)

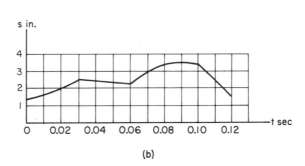

(b)

Figure P8-29

8-30. A cam causes its follower to have the motion shown in Fig. P8-30. Use the graph to plot the v-t and a-t graphs for the follower.

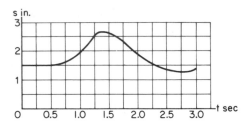

Figure P8-30

8-31. The v-t graph for a particle starts at zero and increases linearly up to 200 ft/sec at 80 sec. It remains constant for the next 30 sec and then decreases linearly to zero at 160 sec. If the initial displacement is -4000 ft, draw the s-t and a-t graphs.

8-32. A v-t graph starts at -40 mi/hr at zero hours and increases linearly to 60 mi/hr at 12 hr. Use graphical procedures to plot the s-t and a-t graphs. The displacement graph starts at zero.

8-33. A particle travels with a constant acceleration of 5 ft/sec² for 10 sec. If its initial speed is -15 ft/sec and its initial displacement is zero, plot the v-t and s-t graphs for the particle.

8-34. A car accelerates at 10 ft/sec² for 8 sec, has no acceleration for 5 sec, then decelerates at 15 ft/sec². Use graphical procedures to determine (a) the length of time the car is in motion if it starts from rest and decelerates until it comes to rest again and (b) the total distance the car travels.

8-35. A particle has a constant acceleration

of a ft/sec², an initial speed of v_1 ft/sec, and an initial displacement of s_1 ft. Use graphical procedures to determine the speed v_2 and position s_2 of the particle after a time interval of Δt.

8-36. A body moves so that its acceleration starts at 40 ft/sec² and decreases constantly to zero at 25 sec. The acceleration remains zero until 45 sec then increases constantly to 15 ft/sec² at 75 sec. If the initial speed is 400 ft/sec and the origin is the starting point, draw the v-t and s-t graphs for the body.

8-37. A body has a constant acceleration of 15 in./sec² for 4 sec, which then decreases linearly to -10 in./sec² during the next 6 sec, and remains constant for the following 8 sec. Plot the v-t and s-t graphs for the particle. Both graphs begin at the origin.

8-38. Sometimes the design of a cam is governed by the acceleration necessary for the follower. The required acceleration for a follower during the part of the cycle after the follower speed is zero is shown in Fig. P8-38. Use this graph to determine the s-t graph for the cam, if s starts at 2 in. The s-t graph obtained will be (approximately) the distance from the center of rotation of the cam to the follower.

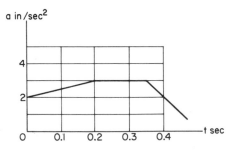

Figure P8-38

8-7. CIRCULAR MOTION OF A PARTICLE

Describing the motion of a particle traveling in a circular path requires somewhat more care than is required for describing the linear motion of a particle, for the direction of the displacement vector is always changing.

Figure 8-17(a) shows the path of a particle traveling in a circle. The position vectors which locate the particle at time t_1 and t_2 are $\mathbf{s}_1$ and $\mathbf{s}_2$ respectively. Remember that the definition for velocity is $\mathbf{v} = \Delta\mathbf{s}/\Delta t$. In this case, $\Delta\mathbf{s}$ is $\mathbf{s}_2 - \mathbf{s}_1$, and $\Delta\mathbf{s}$ can be obtained graphically as shown in Fig. 8-17(b). If $\Delta\mathbf{s}$ is plotted on the path in Fig. 8-17(a), it would join the two points on the path as shown by the dashed line. The velocity will then be the vector $\Delta\mathbf{s}$ divided by the time interval Δt or $t_2 - t_1$. The effect of dividing a vector by a scalar is to change the length of the vector in proportion to the divisor. We will now have a velocity vector which looks something like that shown in Fig. 8-18(a). If we consider the problem for the case where θ is a very small angle, so that Δt is a very small time interval, the velocity vector will change somewhat in position and perhaps in length. When the positions represented by times t_1 and t_2 are so close together that they practically coincide, the velocity vector will be tangent to the path as shown in Fig. 8-18(b), and the velocity obtained will be the instantaneous velocity.

We may then conclude that the instantaneous velocity of a particle traveling in a circular path has a direction tangent to the path at the point where the velocity is being evaluated.

The magnitude of the velocity, obtained from Eq. (8-1), is

$$v = \frac{\Delta s}{\Delta t}$$

For small angles, the length of arc is equal to Δs. Since the arc length is $r(\Delta\theta)$, where r is the radius of the curve, then

$$\Delta s = r(\Delta\theta)$$

and

$$s = \sum (\Delta s) = \sum r(\Delta\theta) = r \sum (\Delta\theta)$$
$$s = r\theta \qquad\qquad (8\text{-}6)$$

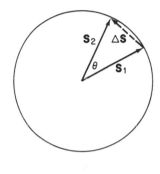

(a)

(b)

Figure 8-17

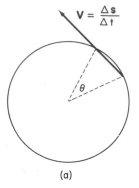

(a)

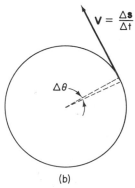

(b)

Figure 8-18

Thus

$$v = r\frac{\Delta\theta}{\Delta t}$$

but

$$\frac{\Delta\theta}{\Delta t} = \omega \qquad\qquad (8\text{-}7)$$

where ω is the rate of change of angle subtended by the path of the particle.

Then $\qquad\qquad v = r\omega \qquad\qquad (8\text{-}8)$

The angle θ is expressed in radians, and the rate of change of angle, ω, better known as the angular velocity, is also expressed in radians per unit of time.

Strictly speaking, both θ and ω are vectors parallel to the axis of rotation and proportional to the magnitude of θ and ω. The vectors are directed according to the right-hand rule. Thus for clockwise rotation, the vector would point into the page (parallel to the -z axis for a plane problem) and would be negative, while for counterclockwise rotation the vector will point out of the page (parallel to the +z axis) and will be positive.

Since the kinematics problems in this text are plane problems, we will be most concerned about the magnitude only of θ and ω along with their proper signs.

EXAMPLE 8-7

A particle is traveling clockwise in a path which is a vertical circle of 3-ft radius. The speed is constant at 75 rpm. Determine the velocity in ft/sec of the particle when it is at the top of the circle, and when it is on the extreme left of the circle.

The speed of the particle must first be calculated using Eq. (8-8) and with ω expressed in rad/sec.

$$\omega = \frac{-75 \times 2\pi}{60} = -7.85 \text{ rad/sec}$$

$$v = r\omega$$
$$= 3 \times (-7.85)$$
$$= -23.55 \text{ ft/sec}$$

The velocity has a direction tangent to the path of the particle.

At the top

$v = 23.55$ ft/sec $\rightarrow$

At the left

$v = 23.55$ ft/sec $\uparrow$

Now that the magnitude and orientation of the velocity has been ascertained, we can determine the acceleration of a particle traveling in a circular path. The velocities v_1 and v_2 corresponding to the times t_1 and t_2 are shown in Fig. 8-19. We will restrict our discussion to the special case where the particle travels around the path with constant speed. In other words, the direction of the velocity will change, but its magnitude will not. From a carefully drawn figure we can find Δv as shown in Fig. 8-19(b). The angle between the two vectors is θ. Note that the direction of Δv is perpendicular to the chord connecting the two velocity vectors in Fig. 8-19(a). Then the acceleration vector, obtained from $a = \Delta v / \Delta t$, will have the same direction so that the acceleration vector, for the case of constant speed, is directed to the center of the circular path.

From Fig. 8-19(b) we can see that

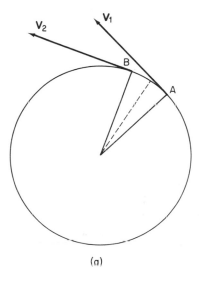

(a)

$$\frac{\Delta v}{2} = v \sin \frac{\theta}{2}$$

or

$$\Delta v = 2v \sin \frac{\theta}{2}$$

For small angles which are measured in radians we have

$$\sin \frac{\theta}{2} = \frac{\theta}{2}$$

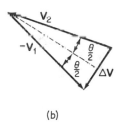

(b)

Figure 8-19

then

$$\Delta v = 2v \frac{\theta}{2} = v\theta$$

For small angles, the chord AB is approximately equal to $r\theta$, where θ is in radians. The distance AB must be equal to $v(\Delta t)$ for small time intervals, since AB is approximately the distance traveled in the time Δt. Then

$$AB = r\theta = v(\Delta t)$$

or

$$v = \frac{r\theta}{\Delta t}$$

and rewriting the above gives

$$\Delta t = \frac{r\theta}{v}$$

Substituting,

$$a = \frac{\Delta v}{\Delta t} = \frac{v\theta}{r\theta/v}$$

$$a_n = \frac{v^2}{r} \qquad\qquad (8\text{-}9)$$

From Eq. (8-8) we have $v = r\omega$. If we substitute this in Eq. (8-9) we have

$$a_n = \frac{v^2}{r} = \frac{(r\omega)^2}{r}$$

$$a_n = r\omega^2 \qquad\qquad (8\text{-}10)$$

Since this acceleration is directed to the center of the circular path, it is normal to the path and is called the normal acceleration. Normal acceleration is the only acceleration a particle has if traveling in a circular path at constant speed.

If the speed of the particle around the circular path is changing, then both the magnitude and direction of the velocity vector will change. That change in velocity will be as shown in Fig. 8-20(a) and will no longer be directed to the center of the circular path. The acceleration vector is then broken up as shown in Fig. 8-20(b), where one component is the normal component a_n and the other component is tangential to the path and is a_t. The magnitude of a_n will be v^2/r. If $\Delta \mathbf{v}$ and Δt can be obtained, the values for $\mathbf{a}$, a_n, and a_t can all be obtained from a vector diagram similar to that shown in Fig. 8-20(b). However, the vector diagram in general provides only a rough approximation of the answer, because the value for Δt is usually not small enough to obtain the instantaneous value for a. If Δt is small enough, then the angle between the two velocity vectors is so small that it is not possible to be sufficiently accurate with the diagram.

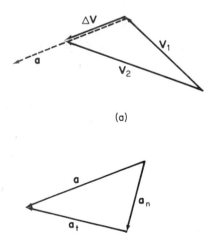

(a)

(b)

Figure 8-20

There will be additional information on obtaining a_t in Section 8-10 on rotational motion of rigid bodies.

EXAMPLE 8-8

Determine the normal acceleration for a particle which is traveling around a 25-ft radius circle at a speed of 100 ft/sec.

The normal acceleration may be obtained directly from Eq. (8-9).

$$a_n = \frac{v^2}{r}$$

$$= \frac{100 \times 100}{25}$$

$$= 400 \text{ ft/sec}^2 \text{ directed to the center of the circle.}$$

PROBLEMS

8-39. A point on the crankshaft of an engine is 3 in. from the center of rotation. If the engine is operating at 3000 rpm, determine the speed, in ft/sec, of the point on the crankshaft.

8-40. A 7-ft diameter flywheel rotates at 25 rpm. Determine the speed of a point on the circumference of the flywheel in ft/sec. If the flywheel is vertical and rotation is counterclockwise, determine the velocity of the point as it passes through the 10 o'clock position.

8-41. A rail diesel car is traveling around a circular curve at 55 mi/hr. If the radius of the curve is 2600 ft, determine the angular velocity of the car in rad/sec.

8-42. A particle travels in a horizontal circle of 18-in. radius at -3π rad/sec. Determine the velocity of the particle when the particle is south 60° east from the center of the path.

8-43. An aircraft is traveling due north and goes into a turn of 3-mi radius so that it ends up on a course due west. Determine the angular velocity in rad/sec of the aircraft in the circular path, if it maintains a constant speed of 350 mi/hr. Also determine the normal acceleration of the aircraft in ft/sec.^{2}

8-44. Find the normal acceleration, in ft/sec^2, for an automobile traveling around the following circular curves (a) 200-ft radius at 20 mi/hr and (b) 5000-ft radius at 70 mi/hr.

8-45. A centrifuge used in a chemistry laboratory operates at 1200 rpm. If the distance from the center of rotation to the sample being separated is 5 in., determine the normal acceleration of the sample.

8-46. The end of a link on a machine must have an acceleration of 4000 ft/sec^2 or less. If the link is 8 in. long, determine the maximum angular velocity, in rpm, that the link can have.

8-47. The typing element used on some electric typewriters rotates through 90° in about 1/200 sec. If the diameter of the element is 1.25 in., determine the normal acceleration of a point on the element, in in./sec^2.

y

S_2

S_1

x

Figure 8-21

8-8. GENERAL PLANE MOTION OF A PARTICLE

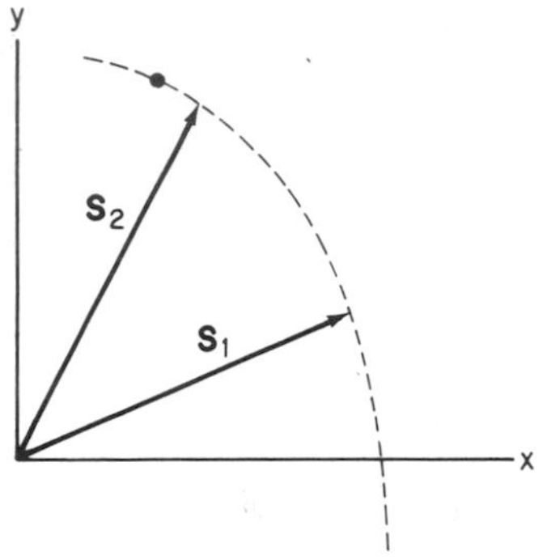

An example of the path of a particle traveling in general plane motion with varying speed is shown in Fig. 8-21. By arguments similar to those used in the discussion of circular motion, it can be shown that the velocity vector will be tangent to the path of the particle.

The acceleration may also be obtained by methods similar to those used in our discussion of circular motion. As with circular motion, the acceleration is also divided into components for convenience. The components used are tangential to the curve and normal to the curve, or radial and transverse, in which case the radial component is along the line from the point where the acceleration is being evaluated to the origin. The transverse component is perpendicular to the radial component. The two types of component are shown in Fig. 8-22.

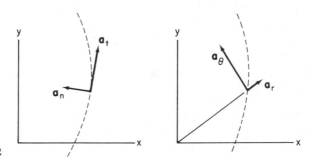

Figure 8-22

A thorough discussion of general plane motion is beyond the scope of this book, although some simple cases will be examined in Sections 8-13 and 8-14.

8-9. RIGID BODY MOTION

A rigid body is a body in which the particles that comprise the body all maintain the same position relative to one another. A nonrigid, or deformable, body is one whose shape and dimensions can change.

For many bodies, such as gears, links, or cams, the shape of the body has an influence on the motion of the body. Thus in order to adequately describe the motion, a knowledge of the kinematics of rigid bodies is required.

Examples of the various types of rigid body motion are shown in Fig. 8-23. A body in rectilinear translation is shown in Fig. 8-23(a) in which all particles move in parallel straight lines. Figure 8-23(b) shows rotational motion in which all particles travel in a circular path about the axis of rotation. In Fig. 8-23(c) the particles that comprise the body do not travel in either a circular or straight-line path. The body is in general plane motion.

For the case of translation, the motion of the body can be adequately described by describing the motion of any particle, for all the other particles must follow in parallel paths. The information necessary to describe rotational motion is given in the next Section. It will then be possible to describe general plane motion after we have discussed relative motion.

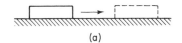

(a)

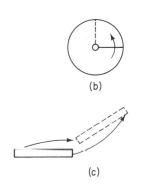

(b)

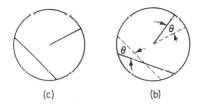

(c)

Figure 8-23

8-10. ROTATIONAL MOTION OF RIGID BODIES

In describing rotational motion, the angular displacement θ is defined in terms of the angle of rotation of a line on the rotating body, as indicated in Fig. 8-24. There are two points that should be noticed about the angular displacement θ. First, it is measured in radians, and it is not limited in size to one revolution or 2π radians. Second, any line on the body would rotate through the same angle as a line which passes through the center of rotation. This is indicated in Fig. 8-24.

The angular velocity ω is the rate of change of the angular displacement, or the rate at which a line on the body rotates. As explained in Section 8-7, Eq. (8-7) is

(c)   (b)

Figure 8-24

$$\omega = \frac{\Delta\theta}{\Delta t} = \frac{\theta_2 - \theta_1}{\Delta t}$$

This gives an average value for angular velocity, except in the case where Δt approaches zero, then the value for ω will be the instantaneous value.

The speed of any point on the line can be obtained from Eq. (8-8)

$$v = r\omega$$

where r is the distance from the center of rotation to the point whose speed is being evaluated. The direction of the velocity will be tangent to the circular path of the point.

The angular acceleration α is the rate of change of the angular velocity

$$\alpha = \frac{\Delta\omega}{\Delta t} = \frac{\omega_2 - \omega_1}{\Delta t} \qquad (8\text{-}11)$$

This gives an average value for angular acceleration, except when Δt approaches zero, in which case the value for α will be the instantaneous value. The angular acceleration is usually measured in rad/sec² or rad/hr². It is a vector, like θ and ω, and the same rules of sign apply.

The tangential acceleration of a particle at a distance r from the center of rotation will be

$$a_t = \frac{r\omega_2 - r\omega_1}{\Delta t} = r\left(\frac{\omega_2 - \omega_1}{\Delta t}\right) \qquad (8\text{-}12)$$

$$a_t = r\alpha$$

Notice that $r\omega$ is the speed, and that a_t is obtained from the change in speed only. This component of acceleration will have the same direction as the velocity, although not necessarily the same sense, which is tangent to the path of the point. This is why the acceleration obtained is called the tangential acceleration.

The total acceleration of a point on a rotating body may be found by taking the vector sum of the normal and tangential components of acceleration from Eq. (8-9) and Eq. (8-12)

$$\mathbf{a} = \mathbf{a}_n + \mathbf{a}_t$$

where the magnitude of $\mathbf{a}$ is

$$a = (a_n^2 + a_t^2)^{1/2}$$

The magnitude may also be obtained using the slide rule method for finding resultants. The direction of the total acceleration may be obtained by determining the angle formed by the two components.

Relationships between time, angular displacement, velocity, and acceleration can be developed using procedures similar to those used to develop relationships for linear motion. The relationships are

$$\omega_2 = \omega_1 + \alpha(\Delta t) \qquad (8\text{-}13)$$

$$\theta_2 = \theta_1 + \omega_1(\Delta t) + \tfrac{1}{2}\alpha(\Delta t)^2 \qquad (8\text{-}14)$$

$$\omega_2^2 - \omega_1^2 = 2\alpha(\theta_2 - \theta_1) \qquad (8\text{-}15)$$

Notice the parallel with Eq. (8-3), (8-4), and (8-5). The subscripts 1 and 2 indicate the initial and final values for displacement and velocity. These relationships are valid only where the angular acceleration α is a constant.

EXAMPLE 8-9

A pulley with a radius of 5 ft has an angular velocity of 5 rad/sec and an angular acceleration of -15 rad/sec^2. Determine the velocity and acceleration of a point on a line 40° from the x axis (a) 3 ft from the center and (b) 5 ft from the center.

From Eq. (8-8) we can determine the speed. The direction will be as shown in Fig. 8-25(a).

(a) $v = r\omega$
$\qquad = 3 \times 5 = 15$ ft/sec

The normal acceleration is obtained using Eq. (8-10).

$a_n = r\omega^2$
$\qquad = 3 \times 5^2 = 75$ ft/sec^2

The tangential acceleration is obtained using Eq. (8-12).

$a_t = r\alpha$
$\qquad = 3 \times (-15) = -45$ ft/sec^2

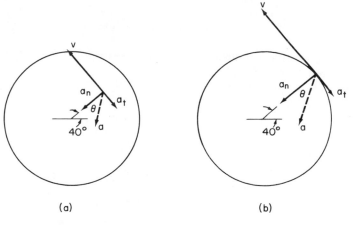

(a)   (b)

Figure 8-25

The total acceleration may be obtained using the Pythagorean theorem or the slide rule method.

$$a = (a_n^2 + a_t^2)^{1/2}$$
$$= (75^2 + 45^2)^{1/2} = (5625 + 2025)^{1/2}$$
$$= 87.5 \text{ ft/sec}^2$$

The angle is the angle between the normal component and resultant and is shown in Fig. 8-25(a).

$$\theta = \tan^{-1}\frac{45}{75} = 31.0°$$

The steps for solving part (b) are the same as for part (a).

(b) $v = r\omega$
$$= 5 \times 5 = 25 \text{ ft/sec}$$
$a_n = r\omega^2$
$$= 5 \times 5^2 = 125 \text{ ft/sec}^2$$
$a_t = r\alpha$
$$= 5 \times (-15) = -75 \text{ ft/sec}^2$$
$a = 145.7 \text{ ft/sec}^2$
$\theta = 31.0°$

EXAMPLE 8-10

A link is rotating at 880 rpm. It is braked so that it decelerates at the rate of 30 rad/sec². Determine (a) the length of time required for the link to come to rest and (b) the number of revolutions that the link will rotate through while coming to rest.

Since the initial and final velocities are known along with α, Eq. (8-13) can be used to find the time required to brake the link to a stop.

(a) $\omega_2 = \omega_1 + \alpha(\Delta t)$

$$0 = \frac{880 \times 2\pi}{60} - 30(\Delta t)$$

$$\Delta t = \frac{880 \times 2\pi}{60 \times 30} = 3.07 \text{ sec}$$

The number of revolutions required to bring the link to rest may be found from either Eq. (8-14) or (8-15). We will use Eq. (8-14) and check our answer using Eq. (8-15).

(b) $\omega_2^2 - \omega_1^2 = 2\alpha(\theta_2 - \theta_1)$

$$0 - \left(\frac{880 \times 2\pi}{60}\right)^2 = -2 \times 30(\theta_2 - 0)$$

$$\theta_2 = \left(\frac{880 \times 2\pi}{60}\right)^2 \times \frac{1}{60}$$
$$= 141.6 \text{ rad}$$

$$\theta_2 = \frac{141.6}{2\pi} = 22.5 \text{ rev}$$

$$\theta_2 = \theta_1 + \omega_1(\Delta t) + \tfrac{1}{2}\alpha(\Delta t)^2$$

$$= 0 + \frac{880 \times 2\pi}{60}(3.07)$$

$$+ \tfrac{1}{2}(-30)(3.07)^2$$
$$= 284 - 142 = 142 \text{ rad}$$
$$\theta_2 = \frac{142}{2\pi} = 22.6 \text{ rev}$$

PROBLEMS

8-48. A gear has a 10-in. diameter and is driven at a speed of 4000 rpm. Determine the speed, in ft/sec, of points 1, 3, and 5 in. from the center.

8-49. A 4-in. diameter pulley must rotate so that the speed of its circumference is 200 ft/min. Determine the angular velocity, in rpm, required for the pulley.

8-50. A link in a mechanism is rotating at 45 rad/sec. If the link is 7 in. long, determine the speed of the end of the link.

8-51. Determine the velocity of a point on a cam 3 in. from the center of rotation and 30° above the positive x axis, if the cam is rotating at 25π rad/sec.

8-52. A flywheel with a radius of 2.5 ft is rotating at 100 rpm and is being accelerated at 50 rad/sec². Determine the speed and the magnitude of the acceleration of a point on the circumference of the flywheel.

8-53. A pulley rotates at -5 rad/sec and has an acceleration of 12 rad/sec². Determine the speed and acceleration of a point 6 in. from the center.

8-54. Find the velocity and acceleration for the end of a link 3 in. long, when the link is in a 4 o'clock position and has an angular velocity of 20 rad/sec and an angular acceleration of 120 rad/sec².

8-55. A gear with an 18-in. diameter must rotate so that the maximum value of acceleration of any point on the gear is 2000 ft/sec². Assuming that the angular acceleration in bringing the gear up to speed is

so small that it can be neglected, determine the maximum angular velocity, in rpm, which the gear may have.

8-56. A point on a rotating body 3 ft from the center of rotation must not accelerate faster than 900 ft/sec². If $\omega = 15$ rad/sec, determine the maximum possible value for α.

8-57. A flywheel on an engine starts from rest and accelerates at 3π rad/sec² until it reaches operating speed of 60π rad/sec. How long does it take for the flywheel to reach operating speed?

8-58. A small electric motor attains its operating speed of 1720 rpm in 3 sec. Assuming constant acceleration, determine the angular acceleration of the motor shaft during starting.

8-59. For each turn of a winch drum a length of cable equal to the circumference of the drum is unwound. If the drum starts from rest and accelerates uniformly at 4π rad/sec², how long will it take to unwind 400 ft from a 3-ft radius drum?

8-60. Determine the time required for a link to rotate through 180° if it starts from rest, accelerates uniformly at 0.3 rad/sec² for the first 90°, and decelerates uniformly at 0.2 rad/sec² for the second 90°.

8-61. A pulley starts from rest and attains an angular velocity of 25 rad/sec after rotating through six revolutions. Determine the angular acceleration.

8-11. GEARS AND BELTS

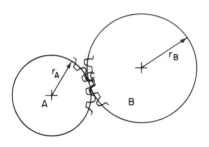

Figure 8-26

Two gears, whose motion is to be analyzed, are represented in Fig. 8-26. Assume that the angular displacement, velocity, and acceleration are known for gear A, and it is necessary to find the angular displacement, velocity and acceleration for gear B.

Notice that as the gears rotate there will always be a common point C. For the brief instant of time that any point is common to both gears, the point must travel through the same distance. From this fact we have

$$s_C = r_A \theta_A$$

and

$$s_C = r_B \theta_B$$

Thus

$$r_A \theta_A = r_B \theta_B$$

or

$$\theta_B = \frac{r_A \theta_A}{r_B}$$

where s, r, and θ represent the distance traveled, the radius, and the angular displacement, respectively, of the points on the gears.

Also, since the point C is common, it has the same speed on both gears so that

$$v_C = r_A \omega_A$$

and

$$v_C = r_B \omega_B$$

Thus

$$r_A \omega_A = r_B \omega_B$$

or

$$\omega_B = \frac{r_A \omega_A}{r_B}$$

where v, r, and ω represent the speed, the radius and the angular velocity respectively, of the points on the gears.

The tangential acceleration for the common point must also be the same for both gears, so that we have

$$a_{t_C} = r_A \alpha_A$$

and

$$a_{t_C} = r_B \alpha_B$$

Thus

$$r_A \alpha_A = r_B \alpha_B$$

or

$$\alpha_B = \frac{r_A \alpha_A}{r_B}$$

where a_t, r, and α represent the tangential acceleration, the radius, and the angular acceleration, respectively, of the points on the gears. Directions may be obtained by observing the direction of speed or acceleration for the common point on each of the gears.

Belt drives are similar to gears except that the motion between pulleys is imparted by a belt instead of the direct contact that exists in gear trains. Since it is assumed that the belt cannot change length, then each point on the belt must have the same speed. Thus if the belt does not slip on the pulleys, the speeds of point C and point D in Fig. 8-27 must both be the same. It can be seen that in effect the belt serves only to separate the common point, so that the relations between angular displacement, velocity, and acceleration for the pulleys are the same as for gears, except for the direction of the motion caused.

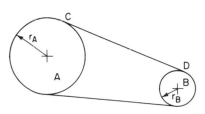

Figure 8-27

EXAMPLE 8-11

Three gears are shown in Fig. 8-28. Gear A is rotating clockwise at 200 rpm. Determine the angular velocity, in rpm, of gear C and the distance traveled by a point on its circumference in 30 sec.

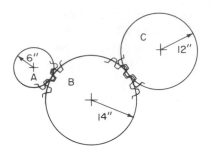

Figure 8-28

The simplest approach is to develop a relationship between the angular motions of the three gears.

$$\omega_B = \frac{r_A \omega_A}{r_B}$$

Similarly

$$\omega_C = \frac{r_B \omega_B}{r_C}$$

Substituting

$$\omega_C = \frac{r_B}{r_C}\left(\frac{r_A \omega_A}{r_B}\right) = \frac{r_A \omega_A}{r_C}$$

Notice that it is not necessary to convert from rpm to rad/sec, since we are dealing only with the ratio of the radii. If A rotates clockwise, then the common point on B must move counterclockwise. The common point on C must then turn clockwise if B turns counterclockwise.

$$\omega_C = \frac{6 \times 200}{12}$$

$$= 100 \text{ rpm clockwise}$$

To find the distance the point travels, θ must be expressed in radians.

$$s_C = r\theta$$

$$= 12 \times \frac{100 \times 2\pi}{60} \times 30$$

$$= 3770 \text{ in.}$$

EXAMPLE 8-12

Pulley A shown in Fig. 8-29 starts from rest and accelerates at 5π rad/sec² for 12 sec. Determine the number of revolutions of pulley B for the 12 sec time interval.

The acceleration of B may be obtained using the ratio of the radii.

$$\alpha_B = \frac{r_A \alpha_A}{r_B}$$

$$= \frac{3 \times 5\pi}{8}$$

$$= 5.89 \text{ rad/sec}^2$$

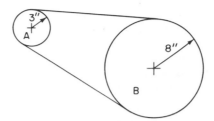

Figure 8-29

The angle of rotation can be obtained using Eq. (8-14). Since the belt causes both pulleys to rotate in the same direction, the value for both α_B and θ must be positive, since α_A was positive.

For B:

$$\theta_2 = \theta_1 + \omega_1(\Delta t) + \tfrac{1}{2}\alpha(\Delta t)^2$$
$$= 0 + 0 + \tfrac{1}{2} \times 5.89 \times 12^2$$
$$= 381 \text{ rad}$$

$$\theta_2 = \frac{381}{2\pi}$$
$$= 60.6 \text{ rev}$$

PROBLEMS

8-62. A gear 5 in. in diameter rotates clockwise at 20 rpm. Determine the angle that the second gear, 12 in. in diameter, rotates through in 1 sec.

8-63. Two gears, with radii of 8 in. and 14 in., respectively, are driven by a gear with a radius of 4 in. If the 4-in. radius gear rotates at 100 rpm, determine the angular velocity of the other two gears.

8-64. If gear D shown in Fig. P8-64 is to rotate at 50 rpm, determine the required angular velocity for the drive gear A.

8-65. Gear A has a radius of 2 in. and is used to drive the train shown in Fig. P8-65. If the angular velocity of B is 4 rad/sec, determine the angular velocity of C. The inside radius of B is 12 in., and the outside radius is 16 in. The radius of C is 5 in.

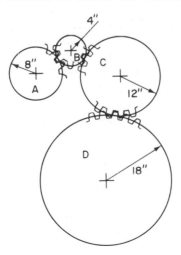

Figure P8-64

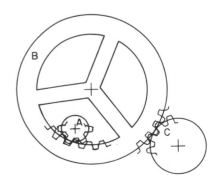

Figure P8-65

8-66. A 0.5 in. radius gear is rotating at 30 rad/sec and drives a second gear of 2-in. radius. If the second gear must rotate through 30 rad in the next 5 sec, determine the necessary acceleration for the drive gear, assuming uniform acceleration.

8-67. A 6-in. radius pulley rotating at 1200 rpm is used to drive a piece of machinery. Determine the size of pulley required for the driven pulley to rotate at 800, 1200, or 1700 rpm.

8-68. An 18-in. diameter drive pulley operates at 300 rpm clockwise. What sizes of pulleys are required to drive a shaft at 150, 200, or 400 rpm.

8-69. A chain is used to connect a 6-in. diameter drive sprocket to a 4-in. diameter sprocket. If the drive sprocket operates at 1500 rpm, determine the speed of the driven sprocket and the size of the sprocket required if it should rotate at 1200 rpm.

8-70. Pulley *A* of the system shown in Fig. P8-70 rotates at 2000 rpm. Determine the speed and direction of rotation of pulley *D*.

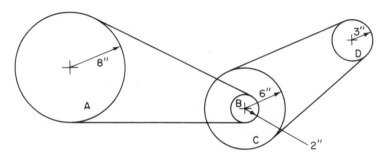

Figure P8-70

8-71. A 3-in. radius pulley starts rotating from rest and accelerates at 3 rad/sec² for 8 sec. Determine the angular velocity of the 7-in. radius pulley being driven and the number of revolutions the 7-in. pulley rotates through in 8 sec.

8-72. A 36-in. diameter drive pulley is rotating at 400 rad/sec. Determine the angular acceleration required if a 16-in. diameter driven pulley is to have a rotation of 200 rad/sec 6 sec after the acceleration begins.

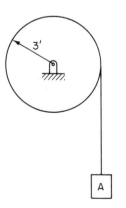

Figure P8-73

8-73. If the rope shown in Fig. P8-73 is wrapped completely around the drum three times, and point *A* is accelerating downward at 4 ft/sec², determine how long it will take to unwind the rope from the drum if it starts from rest.

8-74. A rope is wrapped $5\frac{1}{2}$ times around the drum shown in Fig. P8-74. If the drum is rotating at 5π rad/sec, and accelerating at $-\pi$ rad/sec² determine how long it will take for point B on the rope, which is just coming into contact with the drum, to move to the point at A where it is just leaving contact with the drum.

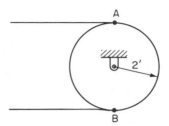

Figure P8-74

8-12. RELATIVE MOTION

A meaningful discussion of even simple cases of general plane motion must be preceded by a discussion of relative motion. First let us consider a few simple general plane problems such as indicated in Fig. 8-30, which shows a ladder sliding down a wall in (a) and a connecting rod BC which connects the piston to the crankshaft as shown in (b). Both of the bodies exhibit general plane motion. However, by means of relative motion relations which describe how one part of a body moves relative to another, it will be possible to analyze and describe the motion of bodies such as the ladder and the connecting rod. Relative motion problems may involve displacement, velocities or accelerations.

Let us first illustrate the concept with a common occurrence, that of one car traveling faster than another on a straight road. Let us assume that car B in Fig. 8-31 is traveling faster than car A. Each car will have a displacement of s_A and s_B, respectively, measured from some common, fixed origin. However, it is also possible, and often convenient, to measure the displacement of car B with respect to car A. The vector from A to B shown in Fig. 8-31(a) is labeled $s_{B/A}$. This vector tells where B is relative to A, as seen by an observer at A. The absolute displacement is the displacement s_A or s_B measured from some fixed point, or as seen by an observer at the fixed point P indicated. At some later time, the displacement of B with respect to A will have a different value as indicated in Fig. 8-31(b).

It should be noted from both figures that a vector equation can be written as follows:

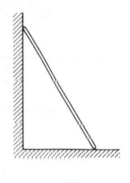

(a)

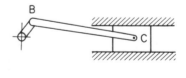

(b)

Figure 8-30

$$s_B = s_A + s_{B/A} \qquad (8\text{-}16)$$

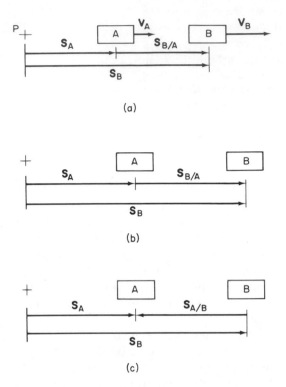

(a)

(b)

Figure 8-31   (c)

The displacement of A with respect to B ($s_{A/B}$) is the opposite of the displacement of B with respect to A ($s_{B/A}$), as indicated in Fig. 8-31(c). Thus we would have

$$\mathbf{s}_A = \mathbf{s}_B + \mathbf{s}_{A/B}$$

Both of the above equations are vector equations, With these equations it is possible to move a term to the opposite side of the equation by changing its sign, as is done with algebraic equations. In manipulating these equations, you will also note that $s_{A/B} = -s_{B/A}$.

In a similar fashion we may determine the relative velocity of B with respect to A. Note that to an occupant of A, car B would appear to be moving ahead, and it would also have a different velocity than it would appear to have to an observer stationed at the edge of the road or point P.

The relationship between the velocity of B and of A may be described as

$$\mathbf{v}_B = \mathbf{v}_A + \mathbf{v}_{B/A} \qquad (8\text{-}17)$$

Similarly, if the velocities of B or A are changing, the relationship between their accelerations may be expressed as

$$\mathbf{a}_B = \mathbf{a}_A + \mathbf{a}_{B/A} \qquad (8\text{-}18)$$

The above relationships may all be manipulated algebraically, as was suggested for the displacement relation. Although all the relations are based on linear motion, they also hold true, as long as expressed in vector form, for general plane motion, or even motion in three dimensions.

Consider the two aircraft A and B shown in Fig. 8-32. The absolute displacements of A and B measured from some arbitrary reference point P are $\mathbf{s}_A$ and $\mathbf{s}_B$. The displacement of B with respect to A is the vector from A to B. The vector equation is then

$$\mathbf{s}_B = \mathbf{s}_A + \mathbf{s}_{B/A} \qquad (8\text{-}16)$$

Similarly, for velocities and accelerations the equations are

$$\mathbf{v}_B = \mathbf{v}_A + \mathbf{v}_{B/A} \qquad (8\text{-}17)$$
$$\mathbf{a}_B = \mathbf{a}_A + \mathbf{a}_{B/A} \qquad (8\text{-}18)$$

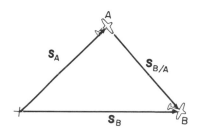

Figure 8-32

EXAMPLE 8-13

Two motorcycles start together on a straight road. If A accelerates at 4.0 ft/sec² and B accelerates at 2.5 ft/sec², determine $\mathbf{a}_{B/A}$, $\mathbf{v}_{B/A}$, and $\mathbf{s}_{B/A}$ after 10 sec.

In order to find $\mathbf{a}_{B/A}$, we may rewrite Eq. (8-18).

Notice that we are dealing with vector equations. Simple algebraic solutions will be correct only if all vectors are parallel. The negative value for the answer indicates that B appears to be accelerating in the opposite direction to that of A.

The velocities at 10 sec may be found from Eq. (8-3).

At 10 sec

$$\mathbf{a}_B = \mathbf{a}_A + \mathbf{a}_{B/A}$$
$$\mathbf{a}_{B/A} = \mathbf{a}_B - \mathbf{a}_A$$
$$= 2.5 - 4.0$$
$$= -1.5 \text{ ft/sec}^2$$

For A

$$v_2 = v_1 + a(\Delta t)$$
$$= 0 + 4.0 \times 10$$
$$= 40 \text{ ft/sec}$$

For *B*

$$v_2 = v_1 + a(\Delta t)$$
$$= 0 + 2.5 \times 10$$
$$= 25 \text{ ft/sec}$$

The value for $v_{B/A}$ may be obtained by rewriting Eq. (8-17).
The negative sign in the answer indicates that *B* appears to be moving backwards with respect to *A*.
The displacement at 10 sec can be obtained from Eq. (8-4).

$$v_B = v_A + v_{B/A}$$
$$v_{B/A} = v_B - v_A$$
$$= 25 - 40$$
$$= -15 \text{ ft/sec}$$

For *A*

$$s_2 = s_1 + v_1(\Delta t) + \tfrac{1}{2}a(\Delta t)^2$$
$$= 0 + 0 + \tfrac{1}{2} \times 4.0 \times 10^2$$
$$= 200 \text{ ft}$$

For *B*

$$s_2 = s_1 + v_1(\Delta t) + \tfrac{1}{2}a(\Delta t)^2$$
$$= 0 + 0 + \tfrac{1}{2} \times 2.5 \times 10^2$$
$$= 125 \text{ ft}$$

We may now obtain $s_{B/A}$ by using Eq. (8-16).

The negative sign in the solution indicates that *B* appears to be behind *A*.

$$s_B = s_A + s_{B/A}$$
$$s_{B/A} = s_B - s_A$$
$$= 125 - 200$$
$$= -75 \text{ ft}$$

EXAMPLE 8-14

A cutting tool moves horizontally to the workpiece shown in Fig. 8-33 at 3 in/sec. If the workpiece is rotating at 20 rpm, determine the velocity at which the point of contact on the workpiece and the cutting tool meet.

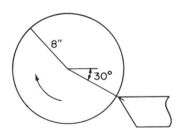

Figure 8-33

The speed of a point on the circumference of the workpiece can be found from Eq. (8-8).

$$v_w = r\omega$$
$$= \frac{2}{3} \times \frac{20 \times 2\pi}{60}$$

$$= 1.398 \text{ ft/sec}$$
$$v_t = 0.25 \text{ ft/sec}$$

The information desired is the relative velocity at which the point on the work-piece and the cutting tool meet. This can be obtained from Eq. (8-17). To solve the vector equation, a sketch should be made, as shown in Fig. 8-34. It might as well be to scale so that a check on the calculations can be made easily. The direction of $\mathbf{v}_t$ was given, and the direction of $\mathbf{v}_w$ must be tangent to the path traveled by the point, which is rotating in a circlar path.

$$\mathbf{v}_t = \mathbf{v}_w + \mathbf{v}_{t/w}$$
$$\mathbf{v}_{t/w} = \mathbf{v}_t - \mathbf{v}_w$$

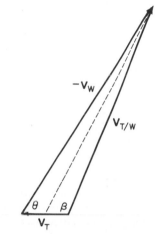

Figure 8-34

The vector triangle can be solved using the cosine law.

$$(v_{t/w})^2 = (v_w)^2 + (v_t)^2 - 2(v_w)(v_t)\cos\theta$$
$$= (1.398)^2 + (0.25)^2$$
$$2 \times 1.398 \times 0.25 \times .5$$
$$= 1.950 + 0.0625 - 0.349$$
$$= 1.663$$
$$v_{t/w} = 1.663^{1/2} = 1.29 \text{ ft/sec}$$

The angle β which gives the orientation of $\mathbf{v}_{t/w}$ can be found using the sine law.

$$\frac{\sin\beta}{v_w} = \frac{\sin 60°}{v_{t/w}}$$
$$\sin\beta = \frac{.866 \times 1.398}{1.29} = .939$$
$$\beta = 110.2°$$

The velocity of contact is 1.29 ft/sec $\angle 69.8°$

PROBLEMS

8-75. Two cars are traveling along a straight highway, with car A traveling at 60 mi/hr and car B traveling at 45 mi/hr. Determine the speed of A with respect to B.

8-76. After impact, two billiard balls continue in a straight line. If ball A has a speed of 2 ft/sec and a speed of 0.5 ft/sec with respect to ball B, determine the speed of ball B.

8-77. An escalator is 40 ft long and travels at 4 ft/sec. If a person tries to go up on the down escalator, what speed must he have relative to the escalator to go up in 13 sec?

8-78. A car is traveling due north at 50 mi/hr and a train is traveling 40° west of north at 70 mi/hr. At one instant as they approach a crossing the train is 3 mi from the crossing, and the car is 2 mi from the crossing. Determine the position of the car relative to the train 2 min later.

8-79. A boat moves across a river at a right angle to the current of 8 mi/hr. If the boat's speed relative to the water is 12 mi/hr, determine the velocity of the boat, and the time it would take to cross the river if the river is 3 mi wide.

8-80. An airplane is to fly a course due east. If its airspeed is 90 mph and the wind

is from north 35° west at 25 mph, determine the course of the aircraft and its ground speed.

8-81. The 36-in. diameter wheel shown in Fig. P8-81 is rotating at 3π rad/sec and accelerating at -8π rad/sec^2 relative to its axle. If the whole assembly is on a carriage accelerating to the left at 8 ft/sec^2, determine the acceleration of A with respect to the ground.

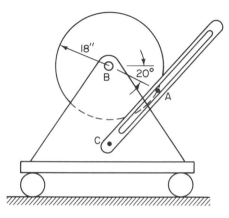

Figure P8-81

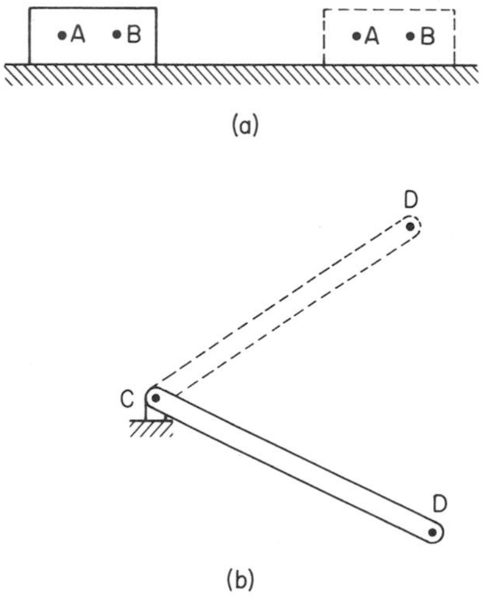

8-13. MOTION IN SYSTEMS OF CONNECTED BODIES

If we examine the motion of particles in rigid bodies a bit more thoroughly, we will get some information necessary to analyze the motion of systems of connected bodies.

Particles which make up a rigid body have some restrictions on the type of motion they may have relative to other particles in the same body. Consider the particles A and B in the body shown in Fig. 8-35(a). Since the body is rigid and moving in linear translation, the particles A and B must remain the same distance from each other. In other words, $s_{B/A}$ is a constant. Since A and B are both in the same body and since $s_{B/A}$ is a constant, then it seems reasonable that $v_{B/A}$ must be zero, since there is no change in displacement. Of course if the velocity of B with respect to A is always zero, the acceleration $a_{B/A}$ must also be zero.

(a)

(b)

Figure 8-35

If a body is rotating about some axis C, the particle at D shown in Fig. 8-35(b) will have to follow a particular path relative to C if the body is a rigid body. In this case, the displacement vector $\mathbf{s}_{D/C}$ will be changing in direction but not in length as the body rotates, for the path of D with respect to C will appear to be a circular path. The velocity of D with respect to C will also change since the displacement is changing. The velocity vector $\mathbf{v}_{D/C}$ will be tangent to the path that D describes relative to C. Similarly, the acceleration of D relative to C, $\mathbf{a}_{D/C}$, can be deduced from the change in $\mathbf{v}_{D/C}$. The components of acceleration will usually be the normal and tangential components. The normal component will always exist, so long as there is a value for $\mathbf{v}_{D/C}$, and the tangential component will exist only if the magnitude of the velocity is changing.

With some understanding of relative motion and the behavior of rigid bodies, it is now possible to discuss problems in which the motion is more general. Let us consider the piston, connecting rod and crankshaft shown in Fig. 8-36(a). If we know the dimensions of the parts and the angular velocity of the crankshaft BC, we can determine the motion of the connecting rod and the piston. For purposes of illustration, let us assume that we want to know the velocity and acceleration of point A which is common to both the connecting rod and the piston. The velocity of A may be found from $\mathbf{v}_A = \mathbf{v}_B + \mathbf{v}_{A/B}$.

This may be worked out vectorially with ease. The velocity $\mathbf{v}_B$ must be perpendicular to BC, since B travels in a circular path about C. The magnitude of $\mathbf{v}_B$ is $r_{BC}\omega_{BC}$. The vector is shown in Fig. 8-36(b). Since A and B are on the same line, and their distance is fixed, the motion of A with respect to B must be circular. Thus the velocity $\mathbf{v}_{A/B}$ must be perpendicular to the line AB, as indicated by the light line in Fig. 8-36(b). We also know that the piston must travel in a straight horizontal line, so that the orientation of $\mathbf{v}_A$ is as indicated by the light horizontal line shown in Fig. 8-36(c). With the construction shown there is only one possible vector triangle which will satisfy the relationship $\mathbf{v}_A = \mathbf{v}_B + \mathbf{v}_{A/B}$. It is shown in Fig. 8-36(d). The value for $\mathbf{v}_A$ may be obtained by graphical solution of the vector triangle or by solving analytically for the sides of the triangle.

The acceleration of A may be calculated in a similar fashion. The acceleration of A may be found from $\mathbf{a}_A = \mathbf{a}_B + \mathbf{a}_{A/B}$.

Assuming that BC has an angular acceleration, the

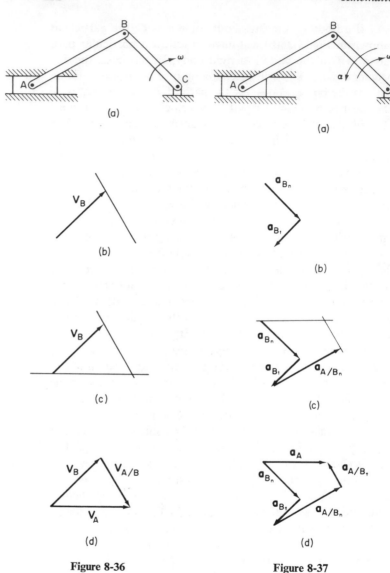

Figure 8-36   Figure 8-37

acceleration of B has both a normal and a tangential component as indicated by the two vectors of Fig. 8-37(b). Since AB
is a rigid body, the motion of A with respect to B must be
circular. If the velocity of A with respect to B has been calculated, then the normal acceleration of A with respect to B can
be calculated and the vector drawn as shown in Fig. 8-37(c).
The tangential component of the acceleration of A with respect
to B can be drawn in, for its direction is known (perpendicular

to *AB*) but its length is unknown. The acceleration of *A* must be horizontal. Thus only one vector diagram can be completed which satisfies the equation $\mathbf{a}_A = \mathbf{a}_B + \mathbf{a}_{A/B}$. This is shown in Fig. 8-37(d).

Emphasis should be placed on the combined use of graphical and analytical solutions. Although these problems can be solved either graphically or analytically, using both procedures provides an excellent check on work at very little expense in time and effort.

EXAMPLE 8-15

The crank *AB* of the slider-crank mechanism shown in Fig. 8-38 has an angular velocity of 3 rad/sec and an angular acceleration of −1 rad/sec². Determine the acceleration of the slider *C*.

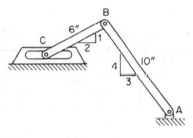

Figure 8-38

To find $\mathbf{a}_C$ we will use the relationship $\mathbf{a}_C = \mathbf{a}_B + \mathbf{a}_{C/B}$. To use the term $\mathbf{a}_{C/B}$, we will need the value for $\omega_{C/B}$ or $v_{C/B}$, which can be obtained only by first finding the velocity. If we find $\mathbf{v}_C$, we can find $\omega_{C/B}$ from $\mathbf{v}_C = \mathbf{v}_B + \mathbf{v}_{C/B}$.

$$\mathbf{v}_C = \mathbf{v}_B + \mathbf{v}_{C/B}$$

$$\mathbf{v}_B = r_{AB}\omega_{AB}$$

$$= 10 \times 3$$

$$= 30 \text{ in./sec}$$

The vector diagram is drawn for the velocity, as shown in Fig. 8-39. The vector $\mathbf{v}_B$ is perpendicular to the line *AB*, and the vector $\mathbf{v}_{C/B}$ must also be perpendicular to the line *BC*. Since the slider must move horizontally, the vector $\mathbf{v}_C$ must go in a horizontal line from the origin to intersect with $\mathbf{v}_{C/B}$. The length of $\mathbf{v}_{C/B}$ can be obtained from the graph or by solving the triangle.

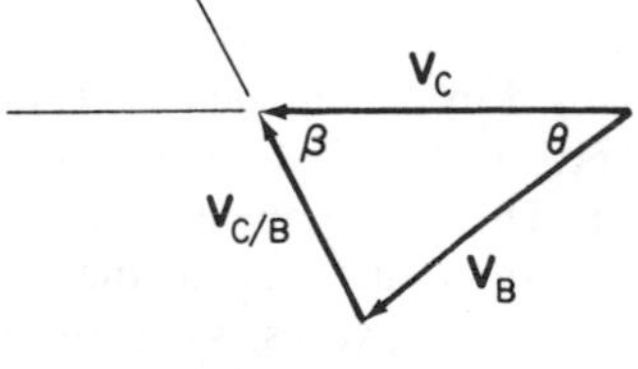

Figure 8-39

$$\theta = \tan^{-1} \tfrac{3}{4} = 36.8°$$

$$\beta = 90° - \tan^{-1} \tfrac{1}{2} = 90° - 26.6°$$

$$= 63.4°$$

$$\frac{v_{C/B}}{\sin \theta} = \frac{v_B}{\sin \beta}$$

$$v_{C/B} = \frac{30 \times \sin 36.8°}{\sin 63.4°}$$

$$= 20.1 \text{ in./sec}$$

Now that we have $v_{C/B}$, we can solve for a_C, using the relative motion relationship.

$$\mathbf{a}_C = \mathbf{a}_B + \mathbf{a}_{C/B}$$

$$a_{B_n} = \frac{v_B^2}{r_{AB}} = \frac{30^2}{10}$$

$$= 90 \text{ in./sec}^2$$

$$a_{B_t} = r_{AB}\alpha_{AB}$$

$$= 10 \times 1$$

$$= 10 \text{ in./sec}^2$$

$$a_{C/B_n} = \frac{(v_{C/B})^2}{r_{BC}} = \frac{20.1^2}{6}$$

$$= 67.1 \text{ in./sec}^2$$

We now have enough information to plot the acceleration vector polygon, as shown in Fig. 8-40. The normal acceleration of B is directed to A along the line of AB, and the tangential acceleration is perpendicular, directed in the direction of the acceleration. The normal acceleration of C with respect to B is directed along the line BC towards B. The tangential acceleration of C with respect to B is unknown, but the direction of its line of action must be perpendicular to the line CB. The acceleration of C must be horizontal and the vector must intersect with the vector for $\mathbf{a}_{C/B_t}$ as shown in Fig. 8-40. The value for $\mathbf{a}_C$ may then be scaled from the diagram. The value for $\mathbf{a}_C$ may also be obtained analytically. First, components are summed in the vertical direction so that $\mathbf{a}_{C/B_t}$ can be found. Then components are summed in the horizontal direction so that $\mathbf{a}_C$ may be found. In both cases, since the vector polygon closes, the sum of the vector components must be such as to equal the resultant vector.

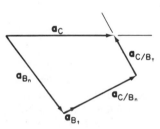

Figure 8-40

$$\sum a_V = 0$$

$$-\frac{4}{5}a_{B_n} + \frac{3}{5}a_{B_t} + \frac{1}{\sqrt{5}}a_{C/B_n}$$

$$+ \frac{2}{\sqrt{5}}a_{C/B_t} = 0$$

$$-0.8 \times 90 + 0.6 \times 10 + 0.445 \times 67.1$$
$$+ 0.890a_{C/B_t} = 0$$

$$a_{C/B_t} = \frac{1}{0.890}(72.0 - 6.0 - 29.8)$$

$$= \frac{36.2}{0.890} = 40.7 \text{ in./sec}^2$$

$$\Sigma \, a_H = 0$$

$$\frac{3}{5} a_{B_n} + \frac{4}{5} a_{B_t} + \frac{2}{\sqrt{5}} a_{C/B_n}$$

$$- \frac{1}{\sqrt{5}} a_{C/B_t} - a_C = 0$$

$$0.6 \times 90 + 0.8 \times 10 + 0.890 \times 67.1$$
$$- 0.445 \times 40.7 - a_C = 0$$

$$a_C = 54.0 + 8.0 + 59.7 - 18.1$$
$$= 103.6 \text{ in./sec}^2 \text{ to the right}$$

8-14. INSTANT CENTER FOR VELOCITY

When considering any rigid body in plane motion there is some point from which the body will appear to have a purely rotational motion at any instant. Consider the ladder sliding down the wall as shown in Fig. 8-41(a). The velocity of point A must be horizontal, and the velocity of point B must be vertical since these points must move as described if contact with the wall and the floor is to be maintained.

For circular motion, the velocity of a point about the axis of rotation is $r\omega$, and the velocity vector is perpendicular to the line from the axis of rotation to the point where the velocity is being evaluated. Thus in Fig. 8-41(a) the perpendiculars to the velocity vectors at A and B intersect at O_1. In other words, O_1 is the only point from which it is possible to draw a line to the point where the velocity is being evaluated and have the velocity vector oriented in its proper direction. This point O_1 is the instant center for velocity at the particular instant of time. The view from O_1 is the same as if A and B were two points on a wheel with center at O_1. At some later time, as shown in Fig. 8-41(b) the new apparent center of rotation will be at O_2, so that O_2 will be the instant center for the ladder in its new position.

In general, the instant center is located by drawing the perpendicular to two known velocities on the body at the points where the velocities are evaluated. The instant center is located at the intersection of the two perpendiculars. If the two velocity vectors are parallel, there can be only one per-

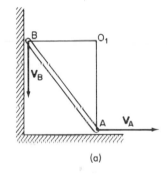

(a)

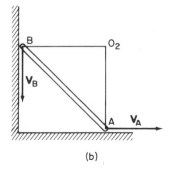

(b)

Figure 8-41

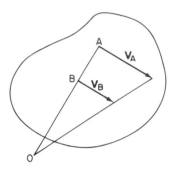

Figure 8-42

pendicular drawn. Because the velocity is proportional to the distance from the instant center, the two velocity vectors are used along with the perpendicular to form similar triangles, which locate the instant center as shown at O in Fig. 8-42.

Once the instant center is located, the velocity of any other point on the body can be obtained by using either similar triangles or the product of the distance to the point and the angular velocity. The direction will be perpendicular to the line from the instant center to the point.

Please note that the instant center for acceleration has a different location than instant center for velocity. In fact, finding the instant center for acceleration is so complex that it is discussed only in very advanced books on mechanics.

EXAMPLE 8-16

The center of the 3-ft diameter wheel shown in Fig. 8-43 moves to the right at 25 ft/sec. Determine the velocities of points B, C, D, and E. The wheel rolls without slipping.

Because several velocities are required, using the instant center will be the most expedient method of solving the problem. To locate the instant center we must first find the velocity of a second point on the wheel. Since the wheel rolls without slipping, the point A must advance a distance equal to the circumference in one revolution of the wheel. Then the speed of any point on the circumference with respect to the center must be the same as the speed of the center.

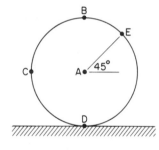

Figure 8-43

The speed of B can be obtained from the relative velocity relation. The velocities of both A and B are parallel, and since the rotation of the wheel is clockwise, both are pointed in the same direction.

$$\mathbf{v}_B = \mathbf{v}_A + \mathbf{v}_{B/A}$$
$$= 25 + 25 = 50 \text{ ft/sec} \rightarrow$$

Now $\mathbf{v}_A$ and $\mathbf{v}_B$ can be drawn as shown in Fig. 8-44. The perpendicular and the line joining the ends of the vectors meet at D. Thus D must be the instant center, and $\mathbf{v}_D = 0$. Similar triangles, as shown in Fig. 8-44 may be constructed to determine

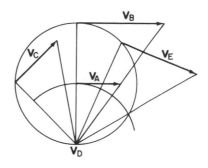

Figure 8-44

the velocities of points C and E. To find v_C, a line is drawn from the instant center D to C, an arc is drawn from A to the line, and v_A is plotted perpendicular to line CD. This is used to form a similar triangle from which v_C is obtained. v_E is obtained in a similar fashion.

By scale

$v_C = 36$ ft/sec $\angle 45°$

$v_E = 48$ ft/sec $\diagdown 22.5°$

$v_D = 0$

PROBLEMS

8-82. The wheel shown in Fig. P8-82 rolls without slipping. If $\omega = 4$ rad/sec, determine the velocity of point B.

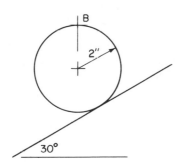

Figure P8-82

8-83. The wheel shown in Fig. P8-83 rolls on its hub without slipping. If the angular velocity of the wheel is 3 rpm clockwise,

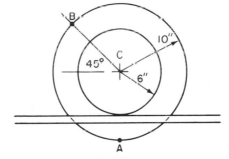

Figure P8-83

determine the velocities of points A and B when the wheel is in the position shown.

8-84. The cylinders shown in Fig. P8-84 have an angular velocity of -3 rad/sec, and their centers are moving with a velocity of 2 ft/sec to the right. If there is no slipping between the plate and the cylinders, determine the velocity of the plate.

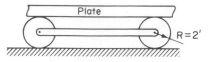

Figure P8-84

8-85. What is the linear velocity of the piston *A* in the position shown in Fig. P8-85 if the crankshaft has a constant angular velocity of 1800 rpm.

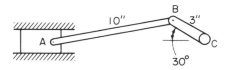

Figure P8-85

8-86. Figure P8-86 shows part of a collapsing structure with a hinge at *B*. If *AB* is 20 ft, *BC* is 10 ft, ω_{AB} is −5 rad/sec, and

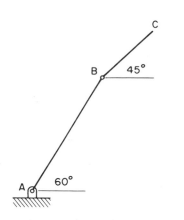

Figure P8-86

ω_{BC} is 2 rad/sec, determine the velocity of *C*.

8-87. If ω for bar *AB* shown in Fig. P8-87 is −5 rad/sec, what is the velocity of point *C*? *B* is a frictionless pin connection and *BC* remains vertical.

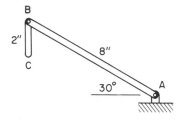

Figure P8-87

8-88. The crank *AB* of the four-bar mechanism shown in Fig. P8-88 rotates at 15 rad/sec. Determine the angular velocity of link *CD*.

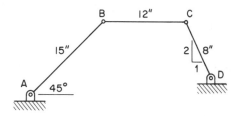

Figure P8-88

8-89. Blocks *A* and *B* are connected by a rigid rod and slide in the tracks shown in Fig. P8-89. At the instant considered, the angular velocity of the rod is 2 rad/sec, and the acceleration of block *A* is 8 ft/sec² to the right. Determine the acceleration of block *B*.

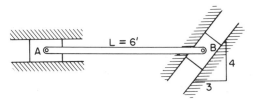

Figure P8-89

8-90. If the wheel shown in Fig. P8-90 rolls on its hub without slipping, determine the acceleration of point B. α is -8 rad/sec^2, and ω is 3 rad/sec.

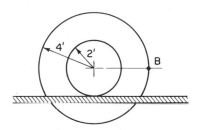

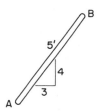

Figure P8-91

P8-91 if a_B is 20 ft/sec^2 to the left and a_A is 6 ft/sec^2 down.

8-92. The acceleration polygon for the mechanism shown in Fig. P8-92(a) is shown in Fig. P8-92(b). Determine the magnitude of the following items using the acceleration polygon: a_B, a_C, a_D, α_{AB} ω_{AB}, and v_B.

Figure P8-90

8-91. Find the angular velocity and angular acceleration of the rod AB shown in Fig.

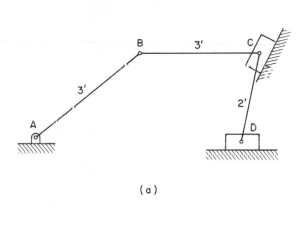

(a)

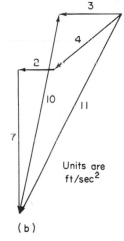

(b)

Figure P8-92

9

KINETICS

Finding out what makes things go should be of interest to anyone with a technical bent. A very important part of making things go is the forces required to cause motion. The study of the relationships between motion of bodies and the forces causing motion is called kinetics. The basis for kinetics is Newton's second law.

It is not sufficient just to make things move, for in most cases the bodies must have a specified motion. Thus, the forces causing motion must be carefully controlled. The tomato harvester shown in Fig. 9 1 requires the application of the principles of kinematics in its design. It takes little imagination to anticipate the condition of the tomatoes if the forces used for picking them are not carefully controlled.

9-1. NEWTON'S SECOND LAW

The whole study of applied mechanics can be reduced in very general terms to the statement of Newton's second law: The change in motion or acceleration of a body is proportional to the resultant force, and in the direction of the resultant force acting on the body. The law can be expressed as

$$\mathbf{a} = k(\sum \mathbf{F})$$

Fig. 9-1. This tomato harvester is one of many places where the principles of mechanics are applied in agriculture. It is a challenge to design a machine which will move easily damaged fruit rapidly and without causing any damage to the fruit. (Photo courtesy of Blackwelder Manufacturing Company.)

where

a is the acceleration,

k is a constant of proportionality, and

$\sum \mathbf{F}$ is the resultant force causing motion.

Experimental results have shown that the value for k is $1/m$ so that the expression may be rewritten as

$$a = \frac{\sum \mathbf{F}}{m}$$

or

$$\sum \mathbf{F} = m\mathbf{a} \qquad\qquad (9\text{-}1)$$

where

$\sum \mathbf{F}$ is the resultant force acting on the body in pounds,
m is the mass of the body in slugs, and
$\mathbf{a}$ is the acceleration in ft/sec².

Since we are working with the English system of measurement, the units listed above will be the only ones used.

The mass of a body is equal to its weight divided by the acceleration due to the gravitational field in which the weight is measured. Near the surface of the earth the acceleration due to gravity, g, is 32.2 ft/sec², so that for weights measured near the earth's surface, the mass is usually about

$$m = \frac{W}{g} = \frac{W}{32.2} \text{ slugs}$$

Note that the mass of a body does not change but is a fixed quantity. The weight of a body may change depending on the attraction due to gravity where the weight is measured.

No book on applied mechanics would be considered complete without a statement of Newton's three laws. Although the second law is the most important, and laws one and three can be deduced from it, for the sake of completeness, laws one and three are as follows. *Newton's first law:* Every body remains at rest or moves with constant velocity in a straight line unless an unbalanced force acts on it. *Newton's third law:* For every action, there is an equal and opposite reaction.

9-2. LINEAR ACCELERATION OF PARTICLES

Before attempting to analyze problems involving acceleration it might be good to recall that the free-body diagram was very useful as an aid to solving equilibrium problems. You will find that a neat, carefully drawn free-body diagram will be of great assistance in solving kinetics problems. Remember that it is very important to show all the external forces acting on the body if the free-body diagram is to be useful. On free-body diagrams for kinetics it is usually a good idea to show the assumed direction of the acceleration with a dashed arrow.

As an example of linear acceleration, consider the block of weight W shown in Fig. 9-2(a). A force of $\mathbf{F}$ is shown acting on the body, and the body is assumed to be on a frictionless surface. The free-body diagram is shown in Fig. 9-2(b). Note that if forces are summed in the y direction and there is no vertical motion then

$$\Sigma F_y = 0$$
$$-W + N = 0$$
$$N = W$$

However, in the x direction we have

$$\Sigma F_x = ma_x$$
$$F = \frac{W}{g} a$$

or

$$a = \frac{Fg}{W}$$

Notice that although $\mathbf{F}$ and $\mathbf{a}$ are vectors, we are following the already established practice of using the scalar magnitudes when working with components in a particular direction.

As another example of linear acceleration, a free-body diagram of a free-falling body in a gravitational field is shown in Fig. 9-3. Notice that the only force acting on the body is the gravitational attraction or the weight of the body. According to Newton's second law, the sum of the forces in the y direction must be

$$\Sigma F_y = ma_y$$
$$-W = \frac{W}{g}(-a)$$

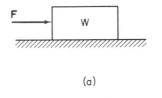

(a)

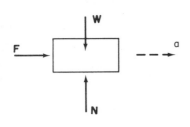

(b)

Figure 9-2

Figure 9-3

then

$$a = g$$

In other words, the acceleration due to gravity of a free-falling body is g if the body is in the earth's gravitational field. The typical value for g is 32.2 ft/sec². Note that in general when we refer to a free-falling body we are neglecting the forces due to air resistance.

One of the reasons for drawing the assumed direction of the acceleration in our free-body diagram is to keep our signs correct. Positive acceleration is up or to the right, and negative acceleration is down or to the left. It is important that the acceleration term be kept on the opposite side of the equation to the force terms. If, in solving the equation, the value obtained for the acceleration is negative, then the assumed direction was opposite to the correct direction.

EXAMPLE 9-1

Determine the acceleration of the block shown in Fig. 9-4(a) if the coefficient of kinetic friction between the block and the surface is 0.30.

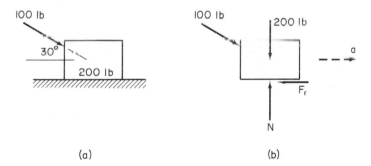

(a)   (b)

Figure 9-4

The first step is to draw the free-body diagram of the block as shown in Fig. 9-4(b).

In order to determine the friction force, we must first find the normal force.

$$\Sigma F_y = 0$$
$$N - 200 - 100 \sin 30° = 0$$
$$N = 200 + 50 = 250 \text{ lb}$$

Then the friction force may be found.

$$F_r = \mu N$$
$$= 0.30 \times 250 = 75 \text{ lb}$$

Now we may determine the acceleration in the x direction by using Newton's second law.

$$\Sigma F_x = ma_x$$
$$100 \cos 30° - 75 = \frac{200}{32.2} a$$

Since the answer has a positive value, our assumed direction for acceleration was correct.

$$a = \frac{32.2}{200}(86.6 - 75)$$
$$= \frac{32.2}{200}(11.6) = 1.87 \text{ ft/sec}^2$$

EXAMPLE 9-2

The rope connecting blocks A and B passes over a smooth rod as shown in Fig. 9-5(a). Determine the acceleration of A and B.

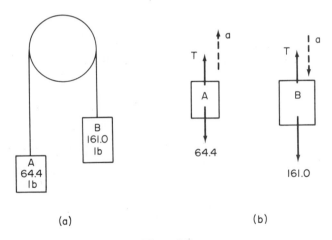

(a)   (b)

Figure 9-5

First draw a free-body diagram of the two blocks as shown in Fig. 9-5(b). Remember that the rope connecting them must have the same tension throughout, since no forces are applied parallel to the axis of the rope except at A and B. For both free-body diagrams we have two unknowns. However, T and a are common to both, so we can set up simultaneous equations to solve for the unknowns.

For A

$$\Sigma F_y = ma_y$$

$$T - 64.4 = \frac{64.4}{32.2}a$$

$$T = 64.4 + 2a$$

For B

$$\sum F_y = ma_y$$

$$T - 161.0 = -\frac{161.0}{32.2}a$$

$$T = 161.0 - 5a$$

Solving simultaneously for a

$$64.4 + 2a = 161.0 - 5a$$

$$7a = 95.6$$

The positive sign indicates that the correct direction was assumed for the acceleration.

$$a = \frac{95.6}{7} = 13.7 \text{ ft/sec}^2$$

PROBLEMS

9-1. A 25-lb force is applied horizontally to a 50-lb weight resting on a smooth horizontal surface. Determine the acceleration of the weight.

9-2. A 644-lb block has a horizontal force of 64.4 lb applied to it. If the block rests on a smooth horizontal surface, determine its acceleration.

9-3. The tractive force between tires and pavement for a 3500-lb automobile is 1000 lb. If we assume that for this problem the automobile may be treated as a particle, determine its acceleration.

9-4. A 10-lb body falls near the earth's surface. If there is a force of 1 lb vertically upward due to air resistance, find the acceleration of the body.

9-5. How many pounds of thrust must be applied by retrorockets to a 600-lb satellite for it to fall at a constant speed of 80 ft/sec in the earth's gravitational field if air resistance is $(v/10)^2$ lb, where v is the speed in ft/sec.

9-6. Determine the apparent weight of a 180-lb man standing on the floor of an elevator if the elevator is accelerating upward at 18 ft/sec².

9-7. If an elevator is accelerating downward at 20 ft/sec², determine the apparent weight of a 170-lb man standing on the floor of the elevator.

9-8. The block shown in Fig. P9-8 weighs 100 lb. Determine the acceleration of the block if the surface on which the block slides is frictionless.

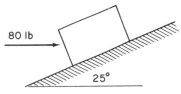

Figure P9-8

9-9. A rope and light pulley are used to move a 400-lb block up an inclined plane as shown in Fig. P9-9. If the coefficient of friction between the block and the plane is 0.30, determine the force required to give the block an acceleration of 4 ft/sec².

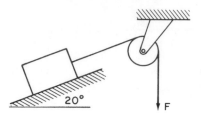

Figure P9-9

9-10. Determine the acceleration of the two blocks shown in Fig. P9-10.

9-11. Determine the force *P* required to

raise the 200-lb weight shown in Fig. P9-11 with an acceleration of 3 ft/sec².

9-12. The weights are released from rest in the position shown in Fig. P9-12. What is the acceleration of block *A*? Block *A* weighs 60 lb, and *B* weighs 100 lb. Ignore pulley weights and axle friction.

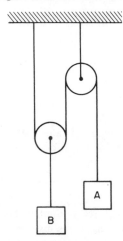

Figure P9-12

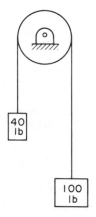

Figure P9-10

9-13. Determine the velocity of body *B* after 5 sec have elapsed if the body starts with a downward velocity of 6 ft/sec. The pulleys for the system shown in Fig. P9-13 are frictionless and of negligible mass.

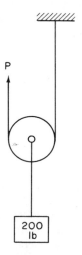

Figure P9-11

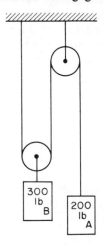

Figure P9-13

9-3. ACCELERATION OF PARTICLES IN A CIRCULAR PATH

According to Newton's first law, a particle cannot travel in a circular path unless a force is exerted on the particle. If a particle is traveling in a horizontal circular path with constant speed as shown in Fig. 9-6(a), there is an acceleration towards the center of the path. The magnitude of the acceleration is $a_n = v^2/r$.

If the particle is of mass m, and we consider the horizontal plane only, then considering the free body of Fig. 9-6(b) we must have

$$\Sigma F_x = ma_x$$

$$F_x = m\frac{v^2}{r}$$

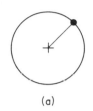

(a)

Thus for the body to move in the prescribed circular path it must have a force exerted on it in the direction of the acceleration, and equal to mv^2/r.

In a discussion of circular motion, reference is often made to *centrifugal* and *centripetal* forces. The centrifugal force is a fictitious force. It is opposite in direction to the acceleration and is equal to ma_n or mv^2/r. The centripetal force is the force exerted on the body to make it follow the circular path. It will also be equal to ma_n or mv^2/r, but will have the same direction as the acceleration, since it is the force that causes acceleration.

If the speed of the particle is also changing then there must be an additional force to cause this acceleration. Such a system is shown in Fig. 9-7. For angular acceleration, the tangential component of acceleration is $a_t = r\alpha$. From the free-body diagram of Fig. 9-7 we have

$$\Sigma F_y = ma_y$$

$$F_y = mr\alpha$$

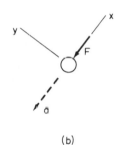

(b)

Figure 9-6

This is the force required to cause the tangential component of acceleration.

EXAMPLE 9-3

A 5-lb weight is on a turntable at a distance of 15 in. from the center of rotation. Determine the total horizontal

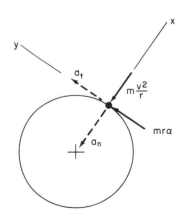

Figure 9-7

force which must exist between the turntable and the weight if there is to be no slipping when the turntable is rotating at 60 rpm and accelerating at 10 rad/sec².

There will be both a normal and a tangential component of acceleration. The forces required to cause each component can be calculated separately and summed vectorially, to find the total horizontal force required to keep the weight in position on the turntable. The necessary free body is shown in plan view in Fig. 9-8.

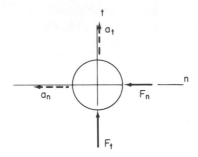

Figure 9-8

Care must be taken to use compatible units, i.e., pounds must be converted to slugs, inches to feet, and rpm to radians per second.

For normal acceleration

$$\Sigma F_n = ma_n$$

$$F_n = \frac{5}{32.2} \times \frac{15}{12} \times \left(\frac{60}{60} \times 2\pi\right)^2$$

$$= 7.65\,\text{lb}$$

For tangential acceleration

$$\Sigma F_t = ma_t$$

$$F_t = \frac{5}{32.2} \times \frac{15}{12} \times 10$$

$$= 1.94\,\text{lb}$$

The slide rule method is used to find the resultant.

Total force = 7.91 lb $\diagdown 14.2°$

PROBLEMS

9-14. An airplane with a speed of 1200 mi/hr flies in a vertical circle. What must be the radius of the curve in order to make a passenger feel weightless at the top of the curve?

9-15. What speed will be required in a car to cause the effect of doubling your weight at the bottom of a vertical curve in a highway if the radius of the vertical curve is 300 ft?

9-16. The operation of some governors is based on the effect of normal acceleration on a weight at the end of a light bar, as shown in Fig. P9-16. If the ball weighs 1.5 lb and the arm AB is 5 in. long, determine the maximum speed (in rpm) allowed if the cutout trips when the angle θ reaches 60°.

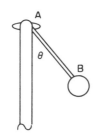

Figure P9-16

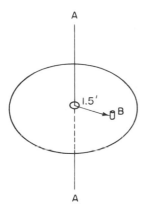

Figure P9-18

9-17. You are to swing a rock weighing 16.1 lb in a 3-ft radius horizontal circle around your head. If the rope breaks with a 20-lb tension, determine the maximum number of rpm at which the rope can be swung.

9-18. The plate shown in Fig. P9-18 rotates about axis *A-A*. If block *B* weighs 3 lb and the coefficient of friction is 0.4, calculate the maximum angular velocity that the plate can have without *B* sliding.

9-19. A 20-lb weight is attached to a 5-ft rope and swings in a vertical arc. Determine the angular acceleration of the rope when it is swinging so that its angle is 30° from the vertical.

9-20. An amusement ride consists of a drum with a 25-ft radius rotating about a vertical axis. Riders stand inside against the wall of the drum as the drum is rotated and the floor is lowered. If the coefficient of friction between the riders' backs and the drum is 0.35, what must be the rate of angular rotation in rpm if the riders are not to slide when the floor is lowered?

9-4. LINEAR MOTION OF BODIES

A rigid body is a collection of particles held together by internal forces. If the rigid body is accelerating in a particular direction in a straight line, it would seem reasonable to expect that each of the particles which make up the body would have the same acceleration in the same direction.

The sum of the accelerating forces on all the particles is such that the acceleration produced is proportional to the total mass of the body and is colinear with the motion of the center of gravity of the body. Consequently, finding the acceleration of a body when the accelerating force acts through the center of gravity of the body is the same as finding the acceleration of a particle. If the resultant accelerating force does not act through the center of gravity of the particle, then the possibility of rotation or tipping must be considered. This will be discussed in Section 9-6.

9-5. ROTATIONAL MOTION OF BODIES

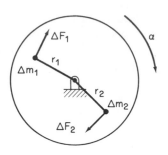

Figure 9-9

If the cylinder shown in Fig. 9-9 is accelerating about its axis of rotation, then each element of mass is being accelerated. The normal acceleration is produced by the internal forces exerted by one element on another. The force required to cause the tangential acceleration of the element of mass Δm is

$$\Delta F = \Delta m\, a_t = r\alpha\, \Delta m$$

The sum of all the forces causing the tangential acceleration is zero, for a body rotating about an axis does not move in a straight line. However, all the forces causing acceleration do represent a torque. The magnitude of the torque for all the elements of the cylinder is

$$T = \sum r\, \Delta F = \sum r(r\alpha\, \Delta m) = \sum r^2\alpha\, \Delta m$$
$$= \alpha \sum r^2\, \Delta m$$

You will recall that the definition for the mass moment of inertia is

$$I = \sum r^2\, \Delta m$$

Thus

$$T = I\alpha \qquad\qquad (9\text{-}2)$$

where

T is the torque causing angular acceleration, measured in foot-pounds,
I is the moment of inertia in slug-ft^2, about the axis of rotation, which is also the axis through the center of gravity, and
α is the angular acceleration measured in rad/sec^2.

The expression relating angular acceleration and torque is really a vector equation $\mathbf{T} = I\boldsymbol{\alpha}$, for both the torque and acceleration are vectors. However, for plane problems the torque and acceleration vectors will be perpendicular to the plane of rotation, and thus we will need to concern ourselves with the magnitude and sign only.

9-6. GENERAL PLANE MOTION OF BODIES

In general, the motion of a body in a plane at any particular instant can be considered to be composed of linear

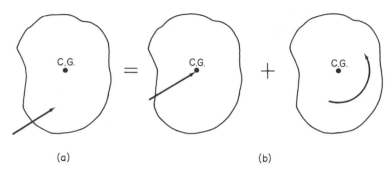

(a)   (b)

translation and rotation. If there is an unbalanced force acting on the body, as shown in Fig. 9-10(a), and if the force does not pass through the center of gravity, the force can be replaced by a force through the center of gravity and a couple, as shown in Fig. 9-10. This is the same procedure which was discussed in Section 3-9. The force through the center of gravity would produce linear acceleration, and the couple would produce angular acceleration. In fact, any system of forces acting on a body can be replaced by a single force through the center of gravity and a couple, in order to simplify acceleration calculations.

Figure 9-10

9-7. D'ALEMBERT'S PRINCIPLE

The solution of a large number of kinetics problems dealing with bodies can be greatly simplified by the use of d'Alembert's principle. According to this principle, kinetics problems may be treated as statics problems if the vector $m\mathbf{a}$ or $I\boldsymbol{\alpha}$ is treated as if it is a force in the opposite direction to $\mathbf{a}$ or $\boldsymbol{\alpha}$. This new "force" is called the *reversed effective force* or *reversed effective couple*. It is also sometimes called the *inertia* of the body.

If the reversed effective force (or couple) is drawn on a free-body diagram, the problem may be solved just like a statics equilibrium problem. In fact, the body represented in the free-body diagram is said to be in a state of *dynamic equilibrium*.

One of the advantages in using the reversed effective force or dynamic equilibrium is that in many cases graphical procedures can be used to solve problems or assist in the solution of problems. Since the body considered is in dynamic equilibrium, a vector polygon of the forces applied, including the reversed effective force, must close. This fact can frequently be used as an aid in finding the magnitude and direction of the acceleration.

For the concept of dynamic equilibrium to be used effectively, there are a few simple rules which must be applied. These will be illustrated in the following discussion and example problems.

An example of a body in linear translation is shown in Fig. 9-11(a). The ordinary free-body diagram, with the assumed

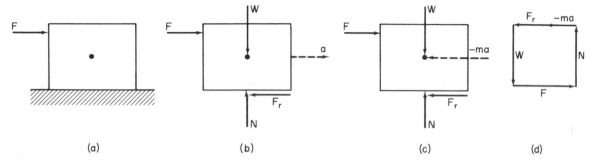

(a)   (b)   (c)   (d)

Figure 9-11   direction of the acceleration, is shown in Fig. 9-11(b), and the free-body diagram for dynamic equilibrium is shown in Fig. 9-11(c). The reversed effective force is shown as a dashed line in the direction opposite to the acceleration. The vector polygon is shown in Fig. 9-11(d) where the vector $-m\mathbf{a}$ is required to close the polygon. In order to obtain the acceleration, it is necessary to divide the $-m\mathbf{a}$ vector by m.

It must be admitted that the value of the vector polygon is marginal in this particular example, for all it really does is confirm the addition of $\mathbf{F} = \mathbf{F}_r + (-m\mathbf{a})$.

For linear translation the reversed effective force must be (1) equal in magnitude to $m\mathbf{a}$, (2) opposite in direction to $\mathbf{a}$, and (3) passing through the center of gravity of the body. For our situation of dynamic equilibrium, the usual equations of equilibrium must be satisfied, that is

$$\Sigma F_x = 0$$
$$\Sigma F_y = 0$$
$$\Sigma M = 0$$

However, it must be kept in mind that $-m\mathbf{a}$ is to be treated as one of the forces on the body.

EXAMPLE 9-4

The file cabinet shown in Fig. 9-12(a) weighs 150 lb. If the coefficient of friction between the cabinet and the floor is 0.30 (static) and 0.20 (kinetic) determine whether the file cabinet will overturn or accelerate. If it will accelerate, determine its acceleration.

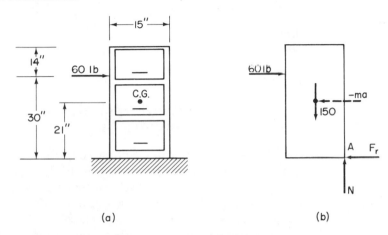

(a)   (b)

Figure 9-12

The first step is to draw a free-body diagram of the file cabinet as shown in Fig. 9-12(b).

The friction force which might exist should first be calculated.

$$\sum F_y = 0$$

$$N - 150 - 0$$

$$N = 150 \text{ lb}$$

$$F_r = \mu N$$

Static friction

$$F_r = 0.30 \times 150 = 45 \text{ lb}$$

Kinetic friction

$$F_r = 0.20 \times 150 = 30 \text{ lb}$$

Accelerating force is large enough to cause motion.

There are three possible actions, sliding, sliding and tipping, and tipping. We have

Assuming sliding

$$\sum F_x = 0$$

already determined that the cabinet will slide, so first we determine its acceleration.

We should then determine whether or not it will tip as it slides. If the cabinet accelerates so that it is on the verge of tipping, the normal reaction will be at corner A, and the acceleration calculated taking moments about A will be that which will just cause tipping. If the actual acceleration is less, the cabinet will slide without tipping.

$$60 - \frac{150}{32.2}a - 30 = 0$$

$$a = \frac{30 \times 32.2}{150} = 6.44 \text{ ft/sec}^2$$

To find the acceleration to cause tipping:

$$\Sigma M_A = 0$$

$$-60 \times \frac{30}{12} + 150 \times \frac{7.5}{12} + \frac{150}{32.2} \times a \times \frac{21}{12} = 0$$

$$8.15a = 150 - 93.7$$

$$a = \frac{56.3}{8.15} = 6.70 \text{ ft/sec}^2$$

File cabinet will accelerate at 6.44 ft/sec² without tipping.

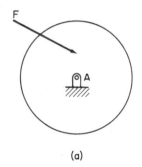

(a)

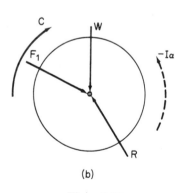

(b)

Figure 9-13

When a body is subjected to a couple only, or a body is restricted to rotational motion such as the flywheel shown in Fig. 9-13(a) in which the flywheel can rotate about the axle A only, we need not be concerned about translation. In order to simplify the analysis, the force F is replaced by the equivalent force F_1 passing through the center of rotation and a couple C. The complete free-body diagram is shown in Fig. 9-13(b). The usual rules for drawing free-body diagrams are observed. However, in order to establish the condition of dynamic equilibrium the appropriate reversed effective force (couple) must be added.

For a body in rotation only, the reversed effective force is a couple equal to $I\alpha$ and opposite in direction to the angular acceleration. All forces except the couple causing acceleration and the reversed effective force (couple) must be concurrent through the axis of rotation. Usually the couple $I\alpha$ is shown as a dashed line to differentiate it from real couples.

The problem may now be treated as a simple moment equilibrium problem for which we must have

$$\Sigma M_A = 0$$

where ΣM_A is the sum of the moments about the axis of rotation, and the reversed effective force (couple) is treated just like any other couple.

EXAMPLE 9-5

The pulley shown in Fig. 9-14(a) weighs 96.6 lb, has a radius of 14 in., and a radius of gyration of 8 in. The belt tensions are as shown. In addition, there is a braking torque, T_A, exerted by the shaft of 25 ft-lb. Determine the angular acceleration of the pulley in rad/sec².

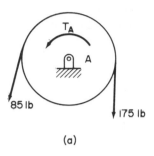

(a)   (b)

Figure 9-14

The first step is to draw a free-body diagram of the pulley, replacing the belt tensions with a force through the shaft and a couple as shown in Fig. 9-14(b). The force R is the reaction from the shaft due to both the weight of the pulley and the tension in the belt.

Torque from the belts

$$T_B = -175 \times \frac{14}{12} + 85 \times \frac{14}{12}$$

$$= -204 + 99.1$$

$$= -104.9 \text{ ft-lb}$$

The moment of inertia is obtained using the radius of gyration.

$$I = k^2 m = \left(\frac{8}{12}\right)^2 \times \frac{96.6}{32.2}$$

$$= 1.33 \text{ slug-ft}^2$$

Moments may now be summed about the axle to determine the acceleration.

$$\Sigma M_A = 0$$

$$-T_B + (-I\alpha) + T_A = 0$$

$$-104.9 + 1.33\alpha + 25 = 0$$

The positive sign for α indicates that the correct direction was assumed. Since $-I\alpha$ was correctly assumed counterclockwise, then α must be clockwise.

$$\alpha = \frac{104.9 - 25}{1.33}$$

$$= 59.8 \text{ rad/sec}^2 \text{ clockwise}$$

A body rotating about an axis not passing through the center of gravity is illustrated in Fig.9-15(a) which shows a bar suspended from point A and accelerating from the position shown. The acceleration of this body can also be determined using d'Alembert's principle. The following reversed effective forces must be included in the free-body diagram as shown in Fig. 9-15(b). Each is shown as a dashed line to differentiate them from true forces or couples.

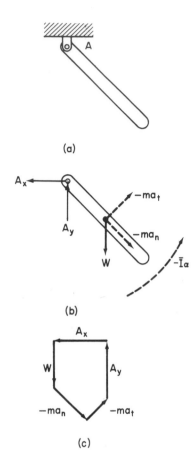

(a)

(b)

(c)

Figure 9-15

1. The reversed effective force due to the normal acceleration of the center of gravity of the body. It is equal to ma_n or $mr\omega^2$, is opposite in direction to a_n, and passes through the center of gravity of the body.
2. The reversed effective force due to the tangential acceleration of the center of gravity of the body. It is equal to ma_t or $mr\alpha$, is opposite in direction to a_t, and passes through the center of gravity.
3. The reversed effective couple due to the angular acceleration of the body. It is equal to $-\bar{I}\alpha$ where $\bar{I}\alpha$ is the moment of inertia about the axis passing through the center of gravity and parallel to the axis of rotation, and α is the angular acceleration. $-\bar{I}\alpha$ is opposite in direction to α.

To treat these problems as dynamic equilibrium problems, the following equations must be satisfied:

$$\Sigma F_x = 0$$
$$\Sigma F_y = 0$$
$$\Sigma M = 0$$

In each case the reversed effective forces are treated as if they were ordinary forces, so that the problems become essentially like equilibrium problems.

The vector polygon of the forces, as shown in Fig. 9-15 (c), can serve as a useful check on the accuracy of the calculations. If a polygon does not close, your answer is wrong.

EXAMPLE 9-6

The uniform bar shown in Fig. 9-16(a) weighs 80 lb and has an angular velocity of -5 rad/sec when in the position shown. Determine the angular acceleration and the reactions at the pin.

First, draw a free-body diagram of the body as shown in Fig. 9-16(b). The reaction at A is most conveniently shown in component form. Notice that the axis system has been oriented parallel to and perpendicular to the bar, since this will require the calculation of fewer components in our solution.

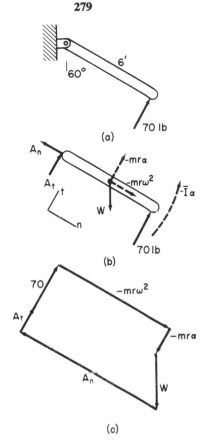

(a)

(b)

(c)

Figure 9-16

We have assumed that the angular accele-ration is negative, hence the reversed effec-tive forces are in the opposite direction.

We will need the moment of inertia, so it might as well be calculated first.

$$\bar{I} = \frac{ml^2}{12} = \frac{80 \times 6^2}{32.2 \times 12}$$

$$= 7.45 \text{ slug-ft}^2$$

In our free-body diagram we have three unknowns, A_n, A_t, and α. If we sum forces in the n direction, we can find A_n directly.

$$\Sigma F_n = 0$$

$$-A_n + 80 \sin 30° + \frac{80}{32.2} \times 3 \times 5^2 = 0$$

$$A_n = 40 + 186$$

$$= 226 \text{ lb}$$

By taking moments about A it is possible to determine the angular acceleration.

$$\Sigma M_A = 0$$

$$-80 \times 3 \cos 30° + \frac{80}{32.2} \times 3\alpha \times 3$$

$$+ 70 \times 6 + 7.45\alpha = 0$$

$$-208 + 22.35\alpha + 420 + 7.45\alpha = 0$$

$$29.80\alpha = -212$$

$$\alpha = -\frac{212}{29.80} = -7.11 \text{ rad/sec}^2$$

Acceleration is counterclockwise

We can now find A_t by summing forces in the t direction. Notice that the negative value for α must be used because the wrong direction was originally assumed.

$$\sum F_t = 0$$

$$A_t + \frac{80}{32.2} \times 3 \times (-7.11)$$

$$- 80 \cos 30° + 70 = 0$$

$$A_t = 53.0 + 69.3 - 70$$

$$= 52.3 \text{ lb}$$

The vector polygon can be used as a check on our work. First, we must determine the lengths of $mr\alpha$ and $mr\omega^2$.

$$mr\alpha = \frac{80}{32.2} \times 3 \times 7.11$$

$$= 53.0 \text{ lb}$$

There is now enough information to draw the vector polygon as shown in Fig. 9-16(c). Since the polygon closes, our answers may be correct, although closure does not guarantee a correct answer.

$$mr\omega^2 = \frac{80}{32.2} \times 3 \times 5^2$$

$$= 186 \text{ lb}$$

PROBLEMS

9-21. An homogeneous block weighing 200 lb is 4 ft high and 3 ft wide. A force of 40 lb is applied horizontally at the top of the block. If the coefficient of friction between the block and the floor is 0.15, determine whether the block will slide or tip, and if it slides, determine the acceleration.

9-22. A 5-ft high crate of uniform density weighing 300 lb is being moved on a dolly. If the coefficient of friction between the crate and the dolly is 0.35, determine the maximum acceleration that the dolly can have without the crate sliding or tipping. The base of the crate is 3 ft × 3 ft.

9-23. The 200-lb homogeneous block shown in Fig. P9-23 is free to slide down the ramp. If the coefficient of friction is 0.20, determine whether the block will tip or slide, and if it slides, determine the acceleration.

9-24. A car weighing W lb is traveling

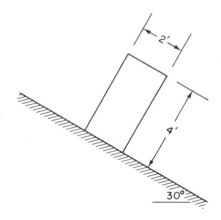

Figure P9-23

around a horizontal curve of radius R. Determine the angle θ of bank or superelevation as shown in Fig. P9-24 required so that the car would not slide off the road at a speed of v ft/sec even if the coefficient of friction is zero.

Figure P9-24

9-25. The homogeneous blocks *A* and *B* shown in Fig. P9-25 weigh 60 lb each. *B* is pin connected to *A* at point *P*. Determine the largest force *Q* that can be applied without causing *B* to tip.

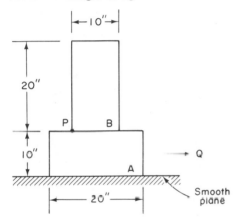

Figure P9-25

9-26. A fork-lift truck weighs 2500 lb and is used to lift a weight of 2200 lb as shown in Fig. P9-26. Determine the upward acceleration of the crate for which the reaction at the rear wheels is zero.

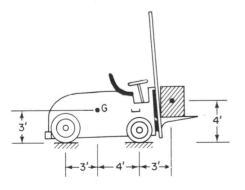

Figure P9-26

9-27. A car traveling on a straight stretch of road decelerates at 10 ft/sec² by locking all four wheels. If the wheelbase of the car is 13 ft, and the center of gravity is 2.5 ft above the ground and 6 ft behind the front wheels, determine the vertical forces on each of the front and rear wheels. The car weighs 3000 lb.

9-28. A bolt located 2 in. from the center of an automobile wheel is tightened by applying the couple shown in Fig. P9-28 to a wrench. Assuming the wheel is free to rotate, determine the angular acceleration of the wheel. The wheel has a radius of gyration of 6 in. and weighs 45 lb.

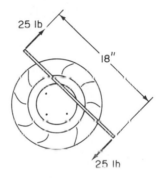

Figure P9-28

9-29. A weight of 100 lb is attached to a cord wrapped around the rim of a flywheel of 4 ft radius. The flywheel weighs 500 lb and has a radius of gyration of 3 ft. Knowing that the system is released from rest, determine the speed of the weight after it has moved down 8 ft. Neglect friction.

9-30. A constant force *Q* of 300 lb acts at an angle of 15° with a tangent to a crank disk *A*, as shown in Fig. P9-30. The radius of gyration of all the rotating parts, including the flywheel *B*, is 2.5 ft. If the speed increases from 60 rpm to 240 rpm during a rotation of 300 revolutions, what is the weight of the rotating parts?

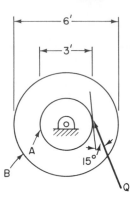

Figure P9-30

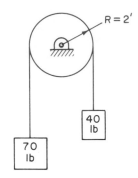

Figure P9-32

9-31. Determine the magnitude of the counterweight *B* necessary to produce an angular acceleration of 12 rad/sec² of the drum shown in Fig. P9-31. The 4-ft diameter drum weighs 20 lb and has a radius of gyration of 1.0 ft.

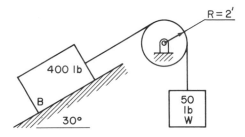

Figure P9-33

9-34. The beam shown in Fig. P9-34 weighs 644 lb and is being lowered from a great height. As it nears the ground, the winch drums are braked so that the deceleration of *A* is 4 ft/sec² and of *B* is 28 ft/sec². What is the tension in each cable?

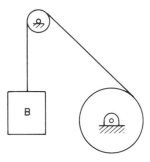

Figure P9-31

9-32. For the system shown in Fig. P9-32, determine the time required for the angular velocity of the drum to reach 5 rad/sec if the system starts from rest and there is no slipping. The drum weighs 200 lb and *k* is 1.2 ft.

9-33. The block *B* and weight *W* shown in Fig. P9-33 accelerate at 2 ft/sec². The pulley weighs 64.4 lb and has a radius of gyration of 0.8 ft. Find the coefficient of friction between the block and the plane.

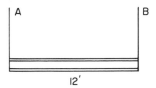

Figure P9-34

9-35. The uniform slender rod shown in Fig. P9-35 is pinned at *A*. When it is released from rest at $\theta = 30°$, what is the angular acceleration?

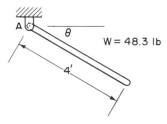

Figure P9-35

9-36. The bar shown in Fig. P9-36 is 4 ft long and weighs 20 lb. The mass at the end weighs 60 lb. Calculate the angular acceleration for the bar for the position shown.

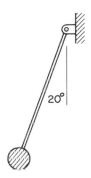

Figure P9-36

9-37. The solid homogeneous cylinder shown in Fig. P9-37 rotates about the axis A. The angular velocity is 4 rad/sec, and

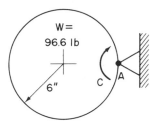

Figure P9-37

the angular acceleration is 12 rad/sec², both counterclockwise. Determine the clockwise couple C applied to the cylinder.

9-38. A rope is wrapped around a solid, homogeneous cylinder as shown in Fig. P9-38. Find the speed of the mass center C after it has dropped 6 ft from rest. The radius of the cylinder is 2 ft.

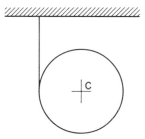

Figure P9-38

9-39. The 100-lb circular cylinder shown in Fig. P9-39 is slipping on the incline. Find the tension in the cord. The coefficient of friction is 0.2. There is a sufficient supply of cord wrapped around the cylinder for movement.

Figure P9-39

10

WORK AND ENERGY

Many problems concerning the motion of bodies and the forces involved in producing motion can be solved using Newton's second law. However, in some instances, solutions using Newton's second law are very tedious, if not practically impossible. Using the methods relating work and energy, which will be discussed in this chapter, it becomes possible to solve many of the problems quite efficiently.

A typical example of a situation where the principles of work and energy are put to use is illustrated in Fig. 10-1. The pile driver shown is rated in terms of the energy it is capable of applying to the pile.

10-1 WORK

For our purposes, work may be defined as the product of the component of force in the direction of the displacement of a body and the displacement of the body.

In Fig. 10-2(a) the work done by the force **F** as the body moves through the displacement **s** is Fs. For Fig. 10-2(b), where the force is not parallel to the displacement, we have

$$U = Fs \cos \theta \qquad (10\text{-}1)$$

Fig. 10-1. This pile driving hammer delivers 19,000 ft-lb of energy per blow, as the 6500 lb ram falls through a 16 in. stroke. The energy comes from gravity and the work done by an hydraulic system. (Photo courtesy of Raymond International Limited.)

where

U is the work done,
F is the magnitude of the force,
s is the magnitude of the displacement, and
θ is the angle between the force and the displacement.

The most commonly used unit of work is the foot-pound although the inch-pound could also be used. The units of foot-pounds will be used most of the time to be compatible with other units which will be encountered.

There is a sign convention associated with work. If the displacement is in the direction of the force, work is positive, and if the displacement is opposite to the direction of the force, the work is negative. The most common example of negative work is that done by a friction force. Since friction opposes relative motion, the friction force will usually be opposite to the displacement, and thus the work done by the friction force will be negative.

For a body moving in the earth's gravitational field there is work done providing the displacement has a vertical component. In Fig. 10-3 the force acting on the body is the gravitational attraction or weight. If the body moves up, the

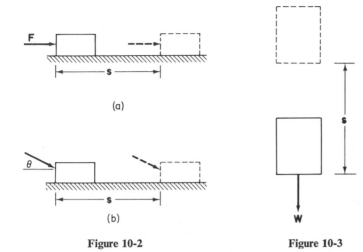

Figure 10-2   Figure 10-3

work done on the body by gravity is going to be $U = -Ws$. If the body moves down, the work done by the force of gravity on the body will be $U = Ws$.

EXAMPLE 10-1

A 200-lb block rests on a level surface. The coefficient of friction between the block and the surface is 0.15. Determine the work done on the block if a horizontal force of 55 lb is applied to the block for a distance of 8 ft.

There will be two horizontal forces doing work on the body. These are the friction force and the 55-lb force. Both will act through the same distance.	
First we shall calculate the friction force.	$F_r = \mu N$ $\quad = 0.15 \times 200$ $\quad = 30\ \text{lb}$
The work done by the friction force, since it opposes motion, as shown in Fig. 10-4, is negative.	$U_1 = Fs$ $\quad = -30 \times 8$ $\quad = -240\ \text{ft-lb}$
The work done by the 55-lb force can also be calculated.	$U_2 = Fs$ $\quad = 55 \times 8$ $\quad = 440\ \text{ft-lb}$

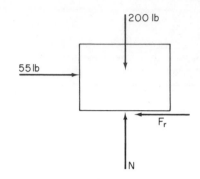

Figure 10-4

The total work is the algebraic sum of the work done.

$$U_{total} = U_1 + U_2$$
$$= -240 + 440$$
$$= 200 \text{ ft-lb}$$

EXAMPLE 10-2

Determine the work done on the 110-lb block shown in Fig. 10-5(a) when it has been raised up the inclined plane a vertical distance of 12 ft. The kinetic coefficient of friction is 0.20.

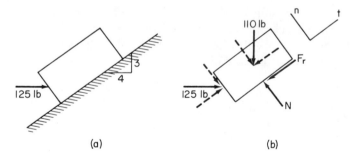

(a)   (b)

Figure 10-5

Draw a free-body diagram of the block as shown in Fig. 10-5(b). Using the free-body diagram, it is possible to calculate the normal force and the friction force. Notice that n and t axes are used for convenience.

$$\sum F_n = 0$$
$$-\tfrac{3}{5} \times 125 - \tfrac{4}{5} \times 110 + N = 0$$
$$N = 75 + 88 = 163 \text{ lb}$$
$$F_r = \mu N$$
$$= 0.20 \times 163$$
$$= 32.6 \text{ lb}$$

The n and t components of the forces have been drawn as dashed lines in the free-body diagram. Notice that the n components do no work since they do not move parallel to the displacement. Positive work is done by the tangential component of the force, and negative work is done by the tangential component of the weight and the friction force. The net work done on the body is the algebraic sum of the work done by the three forces.

The displacement along the plane is

$$s = \tfrac{5}{3} \times 12 = 20 \text{ ft}$$
$$U = \tfrac{4}{5} \times 125 \times 20 - \tfrac{3}{5} \times 110 \times 20$$
$$- 32.6 \times 20$$
$$= 2000 - 1320 - 652$$
$$= 28 \text{ ft-lb}$$

PROBLEMS

10-1. A horizontal force of 40 lb is applied to a 90-lb block resting on a smooth level surface. Determine the work done as the block is moved through a distance of 16 ft.

10-2. Two horizontal forces are applied to a body resting on a smooth level surface. If one force is 25 lb to the left and the other is 30 lb to the right, determine the work done on the body as it moves through a distance of 8 ft.

10-3. The block shown in Fig. P10-3 weighs 60 lb and rests on a smooth horizontal surface. Determine the work done by the force if the block moves a distance of 3 ft.

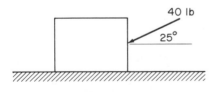

Figure P10-3

10-4. The 25-lb force is applied to the 200-lb block as shown in Fig. P10-4. Determine the work done by the force if the block is moved through a distance of 5 ft.

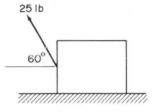

Figure P10-4

10-5. A horizontal force of 25 lb is applied to a block weighing 80 lb. If the coefficient of friction between the block and the horizontal surface is 0.20, determine the work done by the force in moving the block 12 ft and the net work done on the block.

10-6. A 90-lb force is applied horizontally to a 130-lb box on a warehouse floor. If the coefficient of friction between box and floor is 0.40, determine the work done by the force in moving the box 20 ft and the net work done on the box in moving it 20 ft.

10-7. Determine the work done by the 2-kip force applied to the crate shown in Fig. P10-7 in moving it 12 ft. The coefficient of friction between the crate and the floor is 0.25. Also determine the net work done on the crate if it weighs 5 kips.

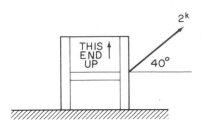

Figure P10-7

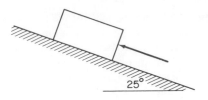

Figure P10-10

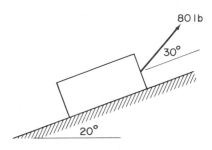

Figure P10-11

10-8. A 50-lb weight is lifted up a stairs through a height of 12 ft. Determine the work that must be done on the body by the lifting force in moving the body through this distance.

10-9. A 50,000-ton ship passing through locks in a canal starts at an elevation of 50 ft in lock *A*, is raised in lock *B* to an elevation of 83.4 ft, and is lowered to an elevation of 48.2 ft in lock *C*. Determine the work which must be done on the ship as it is raised and lowered from lock *A* to lock *C*.

10-10. A 200-lb box is pushed 30 ft along a ramp by a 150-lb force as shown in Fig. P10-10. Determine the work done by the 150-lb force and the net work done on the body. The coefficient of friction is 0.15.

10-11. Determine the work done by the 80-lb force and the net work done on the 60-lb body as it is moved 15 ft along the inclined plane shown in Fig. P10-11.

The coefficient of friction between the body and the plane is 0.25.

10-12. Find the work done on the 160-lb body shown in Fig. P10-12 if it moves a distance of 5 ft along the plane. The coefficient of friction is 0.20.

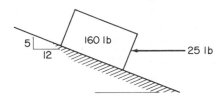

Figure P10-12

10-2. WORK DONE BY A VARIABLE FORCE

Sometimes it is necessary to determine the work done by a variable force such as a spring, or a gas exploding in the cylinder of an engine. The force usually varies in some manner which is related to the displacement. A force-displacement (*F-s*) diagram for such a force is shown in Fig. 10-6, where the force is shown along the vertical axis and the displacement is shown along the horizontal axis. The force shown in the diagram must be the component of the force parallel to the displacement.

The work done during a small displacement Δs is $F_i \Delta s$, the area of the small rectangle shown in Fig. 10-6. The total work done from s_1 to s_2 is $\sum F_i \Delta s$ which would be the sum of a series of rectangles from s_1 to s_2. This is also equal to the area under the F-s diagram from s_1 to s_2. Notice that if the F-s curve falls below the s axis, as occurs at the right end of Fig. 10-6, the work done in this region will be negative.

The most common application of work done by a variable force is in springs. Most springs are linear springs in which the force in the spring (sometimes called the restoring force) is proportional to the stretch of the spring. The force in a linear spring is

$$F = kx \qquad (10\text{-}2)$$

where

F is the force exerted by the spring in lb,
k is the spring constant in pounds per inch, and
x is the deformation, in inches, of the spring, *measured from its unstretched length.*

Thus the force in a spring can be shown graphically as indicated in Fig. 10-7, where the force on the spring will be compression if the spring is being shortened and will be tension if the spring is being lengthened from its unstretched condition.

The work done in stretching a spring some length x_1 from its unstretched position can be calculated from the shaded area shown in Fig. 10-8(a). The work done is the area of the triangular shaded portion, which is $\frac{1}{2} F_1 x_1$. However, the force F_1 is kx_1 for a linear spring, so that the work done may now be written as

$$U = \tfrac{1}{2}F_1 x_1 = \tfrac{1}{2}(kx_1)x_1 = \tfrac{1}{2}kx_1^2$$

As illustrated in Fig. 10-8(b), we can also determine the work done in stretching the spring from x_1 to x_2. Again the

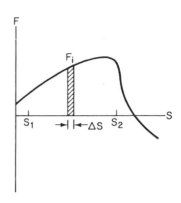

Figure 10-6

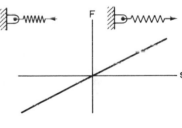

Figure 10-7

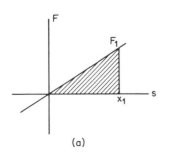

(a)

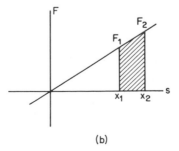

(b)

Figure 10-8

work done may be computed by calculating the area under the graph from x_1 to x_2. In this case we have

$$U = \tfrac{1}{2}(x_2 - x_1)(F_2 + F_1)$$
$$= \tfrac{1}{2}(x_2 - x_1)(kx_2 + kx_1) \qquad (10\text{-}3)$$
$$U = \tfrac{1}{2}k(x_2^2 - x_1^2)$$

The sign convention for work done by a variable force is the same as the sign convention used previously for work. A force which stretches a spring is moving in the same direction as the displacement and is thus doing positive work. The spring itself would be doing negative work in this case, for the force in the spring would be opposite to the direction of the displacement.

EXAMPLE 10-3

The force exerted by a piston on a connecting rod in a full cycle is shown in Fig. 10-9. Determine the net work done by the piston in one cycle.

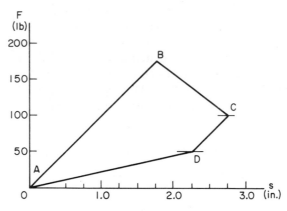

Figure 10-9

The work done in the first half of the cycle will be equal to the area under the curve *ABC*; the work done during the second half will be equal to the area under *CDA*. Since the direction of motion will be reversed during the second half of the cycle, the work done under *CDA* will be negative.

The figure is broken up into triangles and trapezoids in order to calculate the work done.

Area under AB

$= \frac{1}{2} \times 1.75 \times 175 = 153.1$ in-lb

Area under BC

$= \frac{1}{2} \times 1.00 \times (175 + 100)$

$= 137.5$ in-lb

Area under CD

$= \frac{1}{2} \times 0.50 \times (100 + 50)$

$= 37.5$ in-lb

Area under AD

$= \frac{1}{2} \times 2.25 \times 50 = 56.3$ in-lb

$U = 153.1 + 137.5 - 37.5 - 56.3$

$= 196.8$ in-lb

EXAMPLE 10-4

A spring with a constant of 4 lb/in. is extended from compression of 2 in. to elongation of 5 in. Determine the work done by the spring during its total change in length.

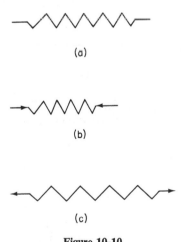

(a)

(b)

(c)

Figure 10-10

The spring in its unstretched position is shown in Fig. 10-10(a). A free-body diagram of part of the spring and the force acting on the spring in order to change its length is shown in Figs. 10-10(b) and (c). The spring is cut at the right end so that the force shown on the right is the force on the spring.

The force whose work is to be investigated is the force in the spring which is opposite in direction to the force on the spring.

Notice that while the spring is changing from a compressed position to an unstretched position, the force in the spring is in the direction of motion. Thus, during this period the spring will be doing positive

In compression

$U = \frac{1}{2}k(x_2^2 - x_1^2)$

$= \frac{1}{2} \times 4(0^2 - 2^2)$

$= -\frac{1}{2} \times 4 \times 4 = -8$ in-lb

work. Following similar reasoning, the work being done by the spring as it is stretched will be negative. These signs are opposite to the answers obtained because Eq. (10-3) is based on work done *on* the spring.

In tension

$$U = \tfrac{1}{2}k(x_2^2 - x_1^2)$$
$$= \tfrac{1}{2} \times 4(5^2 - 0^2)$$
$$= \tfrac{1}{2} \times 4 \times 25 = 50 \text{ in-lb}$$
$$U_{net} = 8 - 50 = -42 \text{ in-lb}$$

PROBLEMS

10-13. The graph shown in Fig. P10-13 shows the force required to stretch a steel rod 1 in. long, and with a cross-sectional area of 1 in.^{2} Determine the work done in stretching the rod 0.002 in.

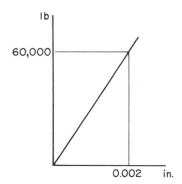

Figure P10-13

10-14. The force exerted on a vehicle in a collision with a fixed object is approxi-

mated by the graph shown in Fig. P10-14. Determine the work done on the vehicle in stopping it.

10-15. The work done by a machine element is illustrated by the graph shown in Fig. P10-15. Determine the work done. (Hint: The area of a quarter circle is $\pi ab/4$ where a is the height and b is the base.)

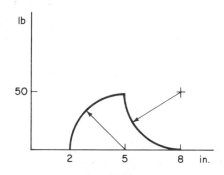

Figure P10-15

10-16. In installing a large piece of machinery, it is moved into place by means of a crane as shown in Fig. P10-16. The machine weighs 20 tons, and the crane pulls with a constant tension of 8 tons. Determine the work done by the crane on the machine as it is moved 30 ft towards the stationary crane.

10-17. Determine the work done by a spring in being stretched 4 in. from its unstretched length if the spring constant is 0.5 lb/in.

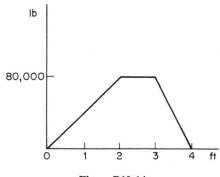

Figure P10-14

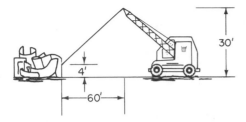

Figure P10-16

10-18. Determine (a) the work done on a spring in stretching it from 2 in. to 5 in. beyond its unstretched length and (b) the work done by the spring when it is being stretched from 1 in. to 4 in. beyond its unstretched length. In both cases the spring constant is 8 lb/in.

10-19. A loaded railway boxcar weighs 160,000 lb and is supported on 8 coil springs in the suspension. In the normal loaded position, the springs are compressed 2 in. Determine the work done on each spring if, after hitting a bump, the springs are compressed to a total of 2.5 in. The spring constant for the springs is 10,000 lb/in.

10-20. The spring shown in Fig. P10-20 is stretched 14 in. Determine the net work done on the weight as the spring retracts to an extended length of 6 in. beyond its unstretched length, if the spring constant is

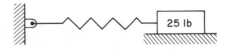

Figure P10-20

7 lb/in. and the coefficient of friction between the block and the floor is 0.25.

10-21. The unstretched length of the spring shown in Fig. P10-21 is 10 in. Determine the work done by the spring on the block in moving the block from the position shown to a point beneath *A*. The weight of the block is 40 lb, the floor is smooth, and the spring constant is 3 lb/in.

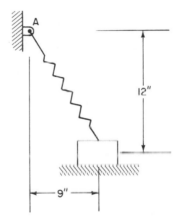

Figure P10-21

10-3. POTENTIAL ENERGY

A body will have potential energy if, due to its position or configuration, it has the potential or ability to do work. In other words, potential energy is energy stored in a body.

A weight *W* a distance *h* above some datum or reference plane, as shown in Fig. 10-11, has potential energy *Wh*, since the work that the weight can do in moving to the datum is *Wh*. In general, then, the potential energy of a body because of its position is

$$V = Wh \qquad (10\text{-}4)$$

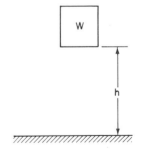

Figure 10-11

where

> V is the potential energy, usually in inch-pounds or foot-pounds,
> W is the weight of the body, and
> h is the height of the body above a datum.

It is sometimes convenient to consider negative potential energy. Here the reference datum is above the body and work must be done on the body to move it to the datum.

A stretched or compressed spring also has potential energy because of its ability to do work. The potential energy stored in a spring is the work done on the spring to put it in its stretched or compressed state. Thus, the potential energy of a spring will be

$$V = \tfrac{1}{2}kx^2 \qquad\qquad (10\text{-}5)$$

where

> V is the potential energy, usually measured in inch-pounds or foot-pounds,
> k is the spring constant, and
> x is the amount the spring has been deformed from its unstretched length.

10-4. KINETIC ENERGY

A body acquires a change in kinetic energy because of the work done on the body to change its motion. The work done on a body is

$$U = Fs$$

but from Eq. (9-1) we have

$$F = ma$$

and from Eq. (8-5) we have

$$a = \frac{v_2^2 - v_1^2}{2s}$$

Thus

$$U = mas = m\left\{\frac{v_2^2 - v_1^2}{2s}\right\}s = \frac{1}{2}\,m(v_2^2 - v_1^2)$$

This represents the change in kinetic energy due to the work done.

If the initial speed of the body was zero, the work done in giving it a speed of v is

$$U = \tfrac{1}{2}mv^2$$

which is the kinetic energy of the body at a speed of v so that we have

$$T = \tfrac{1}{2}mv^2 \qquad (10\text{-}6)$$

where

T is the kinetic energy in foot-pounds,
m is the mass of the body in slugs, and
v is the speed in feet per second.

Since the speed is squared, the kinetic energy of a body will always be positive, although the change in kinetic energy may be positive or negative depending on whether the work done on the body is positive or negative.

Please notice that the above discussion applies to bodies in translation only. The case of work and kinetic energy for rotating bodies will be discussed in Sections 10-6 and 10-7.

10-5. CONSERVATION OF ENERGY

In the previous sections we have discussed how the potential energy and the kinetic energy of a body are both directly related to the work done on the body. Since potential energy is a measure of a body's ability to do work, and kinetic energy is a measure of the work done on a body, it seems logical that the two should be interrelated. The relationship is the law of conservation of energy, which states that the mechanical energy of a body is maintained although it may change its form from kinetic to potential, or vice versa. Expressed as an equation we have

$$V_1 + T_1 = V_2 + T_2 \qquad (10\text{-}7)$$

where

V_1 and V_2 are the initial and final potential energies respectively, and
T_1 and T_2 are the initial and final kinetic energies respectively.
Both V and T must be expressed in the same units.

In many cases the final kinetic or potential energies exist because of work done on the body. The conservation of energy equation is thus modified to take the work into account as follows:

$$V_1 + T_1 + U = V_2 + T_2 \qquad (10\text{-}8)$$

where U is the work done on the system.

The work could be negative if it is work done by a force such as a friction force which removes work from the system. The work done by gravity would not be part of the U term of Eq. (10-8) for it is already incorporated as part of the potential energy term.

The most common use of the principle of conservation of energy is in solving motion problems where the use of Newton's second law directly is time consuming or very inconvenient. Frequently the conservation of energy principle can save several steps compared to a solution involving Newton's second law, particularly if the acceleration of the body is not necessary information.

A simple application of the conservation of energy principle is the calculation of the speed at which a body of weight W strikes the ground after being dropped from a height h. If the ground is taken as our reference point for potential energy we have

$$V_1 + T_1 = V_2 + T_2$$
$$Wh + 0 = 0 + \frac{1}{2}\frac{W}{g}v^2$$
$$v^2 = \frac{2gWh}{W}$$
$$v = \sqrt{2gh}$$

where v is the speed at which the body strikes the ground.

Before beginning work on problems, remember you must be consistent. If you are considering the work done *on* a system, make certain that you do not also include the work done *by* the system, and vice versa.

EXAMPLE 10-5

Determine the speed of the 100-lb block shown in Fig. 10-12(a) after the 80-lb force shown has pushed it a distance of 8 ft along the plane. The kinetic coefficient of friction is 0.25.

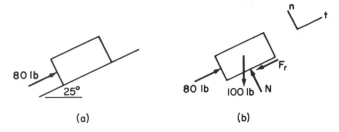

Figure 10-12

First draw a free-body diagram of the block as shown in Fig. 10-12(b) in order to determine the size of the friction force.

$$\Sigma F_n = 0$$
$$N - 100 \cos 25° = 0$$
$$N = 90.6 \text{ lb}$$
$$F_r = \mu N$$
$$= 0.25 \times 90.6$$
$$= 22.7 \text{ lb}$$

Since work is to be done on the body by two forces which will change both the potential and kinetic energy of the body, we will use Eq. (10-8). All the information for this equation is available except for the final speed of the body.

If we use the starting position as our datum, the body will have no initial potential energy, and since it has no initial speed, its initial kinetic energy will also be zero. The 80-lb force does positive work, and the friction force does negative work. The change in potential energy depends on the height to which the body is raised, and the change in kinetic energy depends on the speed. Since the speed is the only unknown, one can readily solve for it.

$$V_1 + T_1 + U = V_2 + T_2$$
$$0 + 0 + 80 \times 8 - 22.7 \times 8$$
$$= 100 \times 8 \sin 25° + \frac{1}{2} \times \frac{100}{32.2} v^2$$
$$\frac{1}{2} \times \frac{100}{32.2} v^2 = 640 - 181.6 - 338$$
$$v^2 = \frac{120.4 \times 2 \times 32.2}{100}$$
$$v^2 = 77.5$$
$$v = 8.80 \text{ ft/sec}$$

EXAMPLE 10-6

A weightless spring as shown in Fig. 10-13 is compressed 2 in. from its free length. The attached 8-lb weight is then released from the position shown. Determine the distance the weight will fall while still attached to the spring. The spring constant is 3 lb/in.

When working with springs, distances must be referred to the unstretched length. Hence the total drop will be $y + 2$ in. as shown in Fig. 10-13.

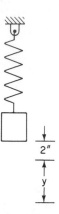

Figure 10-13

The principle of conservation of energy can be used to determine the distance the weight will drop. Since the system starts from rest, there is no initial kinetic energy. However, both the weight and the spring have an initial potential energy. At the end of its travel, the spring will just bring the weight to a halt, so that the weight will have neither potential nor kinetic energy, and the spring will have potential energy.

$V_1 + T_1 = V_2 + T_2$

$\frac{1}{2} \times 3 \times 2^2 + 8(y + 2) = \frac{1}{2} \times 3 \times y^2$

$6 + 8y + 16 = 1.5y^2$

$1.5y^2 - 8y - 22 = 0$

$y = \dfrac{-b \pm \sqrt{b^2 - 4ac}}{2a}$

$\quad = \dfrac{8 \pm \sqrt{8^2 + 4 \times 1.5 \times 22}}{2 \times 1.5}$

$\quad = \dfrac{8 \pm \sqrt{64 + 132}}{3}$

$\quad = \dfrac{8 \pm 14.0}{3}$

Only the positive value of y is a possible solution

$y = \dfrac{8 + 14.0}{3}$

$\quad = 7.33$ in.

Weight drops a total of $7.33 + 2 = 9.33$ in.

PROBLEMS

10-22. A stone is dropped 60 ft from the top of a building. Determine the speed at which the stone strikes the ground.

10-23. A bullet weighing 1 oz is fired ver- tically up from a gun. If the muzzle velo- city is 1200 ft/sec, determine the height that the bullet will reach, neglecting the effect of air resistance.

10-24. A particle weighing 3.22 lb, as shown in Fig. P10-24, is attracted by a magnetic force which varies with the distance from the magnet according to the relation $F = 12/x^2$, where F is in pounds and x is in feet. If the particle has a speed of 10 ft/sec when it is 2 ft from the magnet, what will its velocity be at 1 ft from the magnet. (Hint: A graphical plot of the force-displacement curve will be very helpful.)

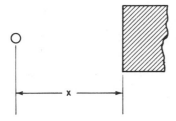

Figure P10-24

10-25. A bumper at the end of a railway track compresses 4 in. while stopping a 40,000-lb boxcar traveling at 3 mi/hr. Determine the spring constant for the spring in the bumper.

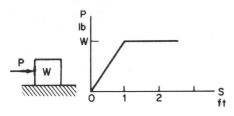

Figure P10-27

10-26. Determine the tractive force required between the pavement and the tires of a 2500-lb car if the car is to reach a speed of 60 mi/hr in a distance of 800 ft after starting from rest.

10-27. A block is accelerated from rest by the variable force P as plotted in the graph shown in Fig. P10-27. What is the velocity of the block after it has travelled 2 ft. There is no friction.

10-28. The block shown in Fig. P10-28 is on a long incline and is released from rest from point A. The block weighs 300 lb and the coefficient of kinetic friction is 0.20. (a) Using the principle of conservation of energy, determine the speed of the block at point B, (b) Using Newton's second law, check your answer to (a).

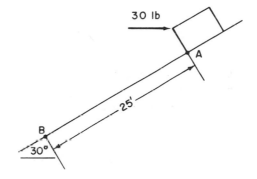

Figure P10-28

10-29. The spring shown in Fig. P10-29 is unstretched in the position shown. Its spring constant is 15 lb/in. If the 12-lb weight shown is just in contact with the spring, determine how far the spring will compress in stopping the weight when it is released.

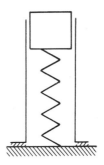

Figure P10-29

10-30. The 6.44-lb collar shown in Fig. P10-30 is pulled out to position A and released from rest. The spring causes motion along the frictionless rod. What is the speed of the collar as it passes position B? The free length of the spring is 6 in., and the spring constant is 4 lb/in.

10-31. The 3-lb weight shown in Fig. P10-31 slides from rest at A along the frictionless rod bent into a quarter ellipse as shown. If the spring has a modulus (spring constant) of 12 lb/ft and an unstretched length of 2 ft, determine the speed of the weight at B.

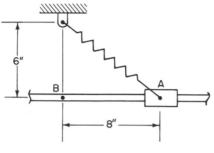

Figure P10-30

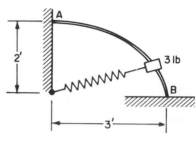

Figure P10-31

10-6. WORK DONE IN CIRCULAR MOTION

A large amount of work-producing devices and equipment rotates. For example, most motors do work through a rotating shaft. The work done by a force rotating through a circular path may be analyzed in the following manner.

Figure 10-14 shows a force applied to a crank rotating about the center O. The distance s that the force travels in rotating from position A to position B is $r\theta$, where θ is measured in radians. From Eq. (10-1) we have $U = Fs$ when the line of action of the force is parallel to its path. However, for this case $s = r\theta$. Thus we have

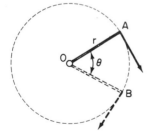

Figure 10-14

$$U = Fs$$
$$= Fr\theta$$

But Fr is equal to the torque produced about the axis of rotation at O, which is T. Thus the work done can also be expressed as

$$U = T\theta \qquad (10\text{-}9)$$

where

U is the work done by the torque,

T is the torque applied, and

θ is the angle (in radians) through which the torque rotates.

If you check the units you will see that the units of work done by a torque are also inch-pounds or foot-pounds depending on the units of the torque.

EXAMPLE 10-7

Determine the work done by an engine in 1 minute if the torque delivered by the driveshaft is 230 ft-lb and the engine is rotating at 600 rpm.

The work can be calculated directly from Eq. (10-8). Care must be taken to convert the number of revolutions in 1 minute to radians.	$\begin{aligned} U &= T\theta \\ &= 230 \times 600 \times 2\pi \\ &= 867{,}000 \text{ ft-lb} \end{aligned}$

10 7. KINETIC ENERGY IN CIRCULAR MOTION

If you have ever tried to stop a rotating wheel, you realize that some effort must be expended. That work done is required to change the kinetic energy of the rotating wheel. The kinetic energy of a rotating body can be determined in the following manner.

The rotating body shown in Fig. 10-15 is made up of many small elements of mass Δm. One of them, placed at a distance r from the center of rotation, is shown in the figure. The kinetic energy of the element of mass is $\frac{1}{2}\Delta m\, v^2$. If the angular velocity of the body is ω, then the kinetic energy of the element may be written as $\Delta T = \frac{1}{2}\Delta m (r\omega)^2$. Thus we have

$$\Delta T = \tfrac{1}{2}\Delta m\, v^2 = \tfrac{1}{2}\Delta m\, r^2\omega^2$$

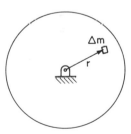

Figure 10-15

The total kinetic energy of the body will be the sum of the kinetic energies of all the elements of mass.

$$T = \Sigma(\Delta T) = \tfrac{1}{2}\Delta m_1 r_1^2\omega_1^2 + \tfrac{1}{2}\Delta m_2 r_2^2\omega_2^2 + \tfrac{1}{2}\Delta m_3 r_3^2\,\omega_3^2 + \ldots$$
$$= \Sigma(\tfrac{1}{2}\Delta m_i\, r_i^2\omega_i^2)$$

Since ω has the same value for any point in the body the expression may be rewritten as

$$T = \tfrac{1}{2}\omega^2 \Sigma\,(r_i^2\,\Delta m_i)$$

From Eq. (7-22) we have

$$I = \sum (r_i^2 \, \Delta m_i)$$

Thus the kinetic energy of a rotating body is

$$T = \tfrac{1}{2}I\omega^2 \qquad\qquad (10\text{-}10)$$

where

T is the kinetic energy, usually in foot-pounds,
I is the mass moment of inertia of the body about the axis of rotation, in slug-ft² or ft-lb-sec², and
ω is the angular velocity in rad/sec.

For a body which is rolling, the value for I will still be about the axis of rotation, which is the instant center for the body.

EXAMPLE 10-8

Determine the kinetic energy of a 2.5-ft diameter truck wheel rolling along the pavement at 30 mi/hr, if the wheel weighs 150 lb and has a radius of gyration of 0.9 ft.

For a body rolling along a surface without slipping, the instant center is the point of contact with the surface, so the moment of inertia must be found with respect to this point.

$$I = mk^2 + md^2$$
$$= \frac{150}{32.2} \times 0.9^2 + \frac{150}{32.2} \times 1.25^2$$
$$= 3.78 + 7.26$$
$$= 10.04 \text{ slug-ft}^2$$

The angular rotation in rad/sec should also be obtained.

$$\omega = \frac{v}{r} = \frac{30 \times 5280}{60 \times 60 \times 1.25}$$
$$= 35.2 \text{ rad/sec}$$

The kinetic energy can now be calculated using Eq. (10-10).

$$T = \tfrac{1}{2}I\omega^2$$
$$= \tfrac{1}{2} \times 10.04 \times 35.2^2$$
$$= 6230 \text{ ft-lb}$$

10-8. CONSERVATION OF ENERGY FOR ROTATIONAL MOTION

The principle of conservation of energy applies whether the motion is linear or rotational. Similarly, work done on a body changes the energy of the body as expressed in Eq. (10-8), and the type of motion (rotation or translation) caused by the work does not affect the validity of the conservation of energy principle.

EXAMPLE 10-9

A 5000-lb flywheel on a punch press rotates at 800 rpm. When some of its energy is used to drive the ram of the press, it slows to 750 rpm. If the ram exerts a force for 1.25 in. of its travel, determine the average force the ram exerts. The radius of gyration of the flywheel is 1.5 ft. The weight of the ram may be neglected.

The rotational kinetic energy of the flywheel is converted to the work done by the ram of the press. The force exerted by the ram may be calculated using the principle of conservation of energy.	
First, calculate the mass moment of inertia of the flywheel.	$I = mk^2$ $= \dfrac{5000}{32.2} \times 1.5^2$ $= 350 \text{ slug-ft}^2$
The conservation of energy equation may now be used. Notice it is rewritten to take into account that some of the initial energy will be used to do work.	$V_1 + T_1 = V_2 + T_2 + U$ $0 + \dfrac{1}{2} \times 350 \left\{ \dfrac{800 \times 2\pi}{60} \right\}^2$ $= 0 + \dfrac{1}{2} \times 350 \left\{ \dfrac{750 \times 2\pi}{60} \right\}^2$ $+ F \times \dfrac{1.25}{12}$ $0.104F = \dfrac{1}{2} \times 350 \left\{ \dfrac{2\pi}{60} \right\}^2 (800^2 - 750^2)$ $= 1.86(640{,}000 - 563{,}000)$ $F = \dfrac{1.86 \times 77{,}000}{0.104}$ $= 1{,}375{,}000 \text{ lb}$

PROBLEMS

10-32. Determine the work done by a small motor in 1 minute if its output torque is 75 in-lb and it operates at 1400 rpm.

10-33. In tightening a nut the last half turn, the force applied at the end of a wrench with a 16-in. handle is 80 lb. Determine the work done, assuming that the force is perpendicular to the wrench handle at all times.

10-34. A brake on a wheel operates in a manner similar to that illustrated in Fig. P10-34. If the force P is 30 lb and the coefficient of friction between the wheel and the brake shoe is 0.25 determine the work done if the wheel is stopped in 15 revolutions.

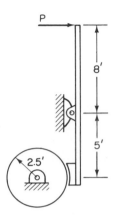

Figure P10-34

10-35. Determine the kinetic energy in an 800-lb flywheel rotating at 1200 rpm. The radius of gyration of the flywheel is 0.8 ft.

10-36. The driveshaft of an automobile is a solid cylinder weighing 150 lb and is 6 ft long and 2.5 in. in diameter. Determine its kinetic energy when it is rotating at 2500 rpm.

10-37. Determine the kinetic energy of a 28-lb slender rod 4 ft long as it rotates about an axis through one end at 15 rad/sec.

10-38. The circular cam shown in Fig. P10-38 weighs 12 lb and rotates about the pin at A at 150 rpm. If it is a uniform disk, determine its kinetic energy.

10-39. In spinning a wheel on a car, a force of 30 lb is applied through an arc of 90°. Determine the angular velocity (in rpm) that the wheel will attain if the wheel weighs 60 lb, has a radius of gyration of 0.6 ft, and a diameter of 2.5 ft.

10-40. A small elevator weighs 800 lb and is raised and lowered by means of a cable over a 2-ft radius drum connected directly to a motor. The combined weight of motor and drum is 600 lb, and their effective radius of gyration is 1.4 ft. Determine the speed of the elevator and the angular velocity of the motor if the elevator is allowed to drop 40 ft without braking.

10-41. A solid, uniform cylinder weighing 240 lb and having a radius of 2 ft starts to roll from rest down an inclined plane. If the plane is inclined at 30°, determine the speed of the cylinder when it has reached the bottom if the height from which the cylinder starts is 4 ft.

10-42. A large reel of cable weighs 1800 lb, has a radius of 3 ft, and a radius of gyration of 2.1 ft. It is to be unwound by means of a man pulling on the cable with a 50-lb force. The reel is supported on a stand which acts as an axle. Determine the speed the man must have to keep the cable running smoothly by the time he is 80 ft from the drum.

10-43. A crane is used to lower a 5-ton crate 40 ft from a ship to the dock. If the hoist drum applies a braking torque of 7500 ft-lb, weighs 4000 lb, and has a radius of gyration of 2.5 ft and a radius of 3 ft, determine the speed at which the crate will strike the ground. It is assumed the descent was started from rest.

Figure P10-38

10-9. POWER AND EFFICIENCY

The rate at which work is done is a measure of the power of a machine capable of doing work. Power may be expressed in several different ways. The most common way is

$$P = \frac{U}{t} \tag{10-11}$$

where
P is the power,
U is the work done, and
t is the time interval during which work is done.
Since $U = Fs$, Eq. (10-11) may be rewritten as

$$P = \frac{Fs}{t} \tag{10-12}$$

where F is the force applied, and s is the distance through which the force is applied. If s changes constantly, Eq. (8-1) may be rewritten as $\Delta s / \Delta t = s/t = v$, so that we may now write Eq. (10-12) as

$$P = Fv \tag{10-13}$$

where v is the speed of the body doing work.

The units used for power will depend on the units used for work, time and force. The most common units are ft-lb/min, ft-lb/sec, and in-lb/sec.

For rotational motion, power is still defined as indicated in Eq. (10-11). However, since $U = T\theta$, it may be expressed as,

$$P = \frac{T\theta}{t} \tag{10-14}$$

where T is the torque applied, and θ is the angle through which it is applied.

If θ changes constantly, Eq. (8-7) may be rewritten as $\Delta\theta / \Delta t = \theta/t = \omega$, so that we may rewrite Eq. (10-14) as

$$P = T\omega \tag{10-15}$$

where ω is the angular velocity. The units will be ft-lb/min, ft-lb/sec, or in-lb/sec depending on the units of $U, T, t,$ and ω.

Two common measures of power have been standardized. They are the horsepower, which is 33,000 ft-lb/min or 550 ft-lb/sec, and the watt, which is 0.737 ft-lb/sec.

Efficiency is the ratio of power output to power input of a machine, and is usually expressed as a percentage.

$$\text{Efficiency} = \frac{P_o}{P_i} \times 100 \qquad (10\text{-}16)$$

A machine with low efficiency wastes power through friction and heat losses, whereas an efficient machine puts out most of the power supplied to it.

EXAMPLE 10-10

To keep a train moving at 40 mi/hr, a locomotive exerts a pull of 15,000 lb. Determine the horsepower used to pull the cars.

We need to find the work done in ft-lb/sec in order to find the horsepower.

First find the distance traveled in 1 second.

$$s = \frac{40 \times 5280}{60 \times 60}$$

$$= 58.7 \text{ ft in 1 sec}$$

The power can be obtained from Eq. (10-12), and this can then be converted to horsepower.

$$P = \frac{Fs}{t}$$

$$= \frac{15,000 \times 58.7}{1}$$

$$= 881,000 \text{ ft-lb/sec}$$

$$P = \frac{881,000}{550}$$

$$= 1600 \text{ horsepower}$$

EXAMPLE 10-11

A motor is supplied with 750 watts of power. If it operates at an efficiency of 92 percent, determine the shaft output in horsepower.

First convert the power input to ft-lb/sec for later ease in converting to horsepower.

$$P_i = 750 \times 0.737$$

$$= 553 \text{ ft-lb/sec}$$

Use the efficiency to determine the output work.

$$\frac{P_o}{P_i} \times 100 = 92$$

$$P_o = 553 \times 0.92$$
$$= 509 \text{ ft-lb/sec}$$

$$P = \frac{509}{550} = 0.926 \text{ horsepower}$$

PROBLEMS

10-44. How many horsepower are required to lift a 2500-lb elevator 800 ft in 1.5 minutes?

10-45. A car exerts a pull of 200 lb on a large trailer as it pulls the trailer along a highway at 50 mi/hr. Determine the horsepower used in towing the trailer.

10-46. A motor has a rated output of 75 in-lb of torque at 1700 rpm. Determine its power in watts and horsepower.

10-47. A motor delivers power to a machine through a belt. If the driven pulley has a radius of 8 in. and is rotating at 600 rpm, determine the horsepower delivered to the machine if the tension in the belt on the tight side is 75 lb and 20 lb on the slack side.

10-48. A 5-hp motor is used to operate a winch to lift a 3500-lb weight 60 ft in 2 min. Assuming that the weight of the winch is small, what is the efficiency of the lifting system?

10-49. The power input in a transmission is 800 ft-lb at 2000 rpm. Determine the efficiency of the transmission if the output is 1500 ft-lb at 950 rpm.

10-50. In order to overcome resistance and maintain a steady speed of 60 mi/hr, a 150-lb force must be applied to an automobile. If the power delivered to the wheels is only 15 percent of the rated power, what horsepower engine must there be in the automobile?

11

IMPULSE AND MOMENTUM

Another method for solving kinetics problems is available if the methods of Newton's second law or work and energy are too cumbersome and time consuming. This third method, which uses a different combination of variables than the other two methods, is the impulse-momentum method.

A nontechnical application of the principles of impulse and momentum, which we will discuss in this chapter, is shown in Fig. 11-1. However, we do not guarantee that a mastery of the principles discussed in Chapter 11 will improve your score at billiards.

11-1. LINEAR IMPULSE AND MOMENTUM

The expressions for linear impulse and momentum are developed by a manipulation of Newton's second law. From Newton's second law we have

$$\sum \mathbf{F} = m\mathbf{a}$$

but $\mathbf{a} = \Delta\mathbf{v}/\Delta t$, which can be substituted so that we have

$$\sum \mathbf{F} = m\frac{\Delta\mathbf{v}}{\Delta t}$$

311

Fig. 11-1. Here is an example of a common device for learning some of the principles of applied mechanics experimentally. (Photo courtesy of Brunswick Corporation.)

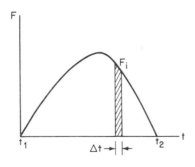

Figure 11-2

If we multiply both sides of the equation by Δt we get

$$\sum \mathbf{F} \, \Delta t = m \frac{\Delta \mathbf{v}}{\Delta t} \Delta t = m \, \Delta \mathbf{v}$$

or

$$\sum \mathbf{F} \, \Delta t = m(\mathbf{v}_2 - \mathbf{v}_1) \tag{11-1}$$

where

$\sum \mathbf{F}$ is the sum of the forces applied to the body,
Δt is the time interval during which the forces are applied,
m is the mass of the body, and
$\mathbf{v}_1$ and $\mathbf{v}_2$ are the initial and final velocities, respectively. The units usually used will be pounds for the forces, seconds for time, slugs for mass, and feet per second for velocity.

The expression $\sum \mathbf{F} \, \Delta t$ is called the linear impulse. It is a vector. If the magnitude of $\mathbf{F}$ varies as shown in Fig. 11-2, the impulse may be calculated by taking the area under the F-t curve. The impulse for a very short time span Δt is the shaded area shown which is $F_i \Delta t$. The total impulse will be the sum of all the small areas, which is the total area under the curve from t_1 to t_2. A variable force as indicated in Fig. 11-2 occurs in many situations, such as a golf club hitting a ball, or a catapult assisting an aircraft in take-off from a carrier.

The expression mv is called the linear momentum. It is also a vector. It is important that this be kept in mind, particularly if the force is not colinear with the initial velocity. The units of both impulse and momentum will be pound-seconds.

EXAMPLE 11-1

A tug exerts a 3000-lb pull for 5 sec on a 500-ton barge in order to get it started moving from dock. Determine the speed of the barge at the end of the 5-sec interval. The frictional force between the water and the barge may be neglected.

Since the force and the velocity must have both the same direction when there are only two vectors to consider, we need to be concerned about the numerical values of the impulse and momentum only. The impulse-momentum equation can be solved directly for v_2.

$$\Sigma F \Delta t = mv_2 - mv_1$$

$$3000 \times 5 = \frac{1,000,000}{32.2} v_2 - 0$$

$$v_2 = \frac{3000 \times 5 \times 32.2}{1,000,000}$$

$$= 0.484 \text{ ft/sec}$$

EXAMPLE 11-2

A 5-oz baseball is traveling horizontally at 60 ft/sec when struck solidly by a bat. If the bat is being swung upward at an angle of 30° with the horizontal, determine the velocity at which the ball will leave the bat if the average force exerted by the bat is 12 lb over a period of 0.17 sec.

We are dealing with a vector equation, so a sketch of the vector solution will be very helpful. With the given information, two of the vectors are known completely, and the third must be solved for.

First determine the magnitude of each vector so that we may use Eq. (11-1) effectively. The known vectors are shown drawn to scale in Fig. 11-3 to satisfy Eq. (11-1). The unknown vector is drawn in as

$$\Sigma F \Delta t = 12 \times 0.17$$

$$= 2.04 \text{ lb-sec}$$

$$mv_1 = \frac{5}{16 \times 32.2} \times 60$$

$$= 0.584 \text{ lb-sec}$$

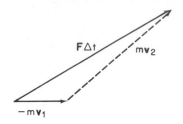

Figure 11-3

a dashed line. Its magnitude and direction can be scaled from the figure.

The vector $m\mathbf{v}_2$ can also be obtained by summing the vertical and horizontal components of the other vectors.

For horizontal components

$$(\textstyle\sum F\,\Delta t)_H = (mv_2)_H - (mv_1)_H$$

$$(mv_2)_H = 2.04\cos 30° - 0.584$$

$$= 1.766 - 0.584$$

$$= 1.182$$

For vertical components

$$(\textstyle\sum F\,\Delta t)_V = (mv_2)_V - (mv_1)_V$$

$$(mv_2)_V = 2.04\sin 30° - 0$$

$$= 1.02$$

$$m\mathbf{v}_2 = 1.560\ \text{lb-sec}\qquad \angle\,40.8°$$

The resultant may be obtained by using the Pythagorean theorem or the slide rule method.

$$v_2 = \frac{mv_2}{m} = \frac{1.560}{\dfrac{5}{16 \times 32.2}}$$

$$= \frac{1.560 \times 16 \times 32.2}{5}$$

The magnitude of the momentum vector must be divided by m to obtain the magnitude of the speed. The direction must be the same as the direction of the final momentum $m\mathbf{v}_2$.

$$\mathbf{v}_2 = 161\ \text{ft/sec}\qquad \angle\,40.8°$$

11-2. CONSERVATION OF LINEAR MOMENTUM

In a system of bodies the momentum of the system will be maintained provided there are no externally applied forces in the direction of motion. This may be deduced using Newton's third law which says that for every action there must be an equal and opposite reaction.

Consider the two skaters shown in Fig. 11-4. Assuming that any friction between the ice and the skates can be neglected, there is no force acting on the skaters parallel to the direction of motion other than the force in the rope which they both hold.

Figure 11-4

The force which A exerts on B must be the same as the force which B exerts on A, and it must also have the same duration. Hence the impulse acting on A must be the same as the impulse acting on B but opposite in direction. Thus when we consider the impulse and momentum for the total system we have

$$\mathbf{F}_A \,\Delta t - \mathbf{F}_B \,\Delta t = (m\mathbf{v}_2)_A - (m\mathbf{v}_1)_A + [(m\mathbf{v}_2)_B - (m\mathbf{v}_1)_B]$$

but

$$\mathbf{F}_A \,\Delta t = \mathbf{F}_B \,\Delta t$$

Thus we have

$$0 = (m\mathbf{v}_2)_A - (m\mathbf{v}_1)_A + (m\mathbf{v}_2)_B - (m\mathbf{v}_1)_B$$

$$(m\mathbf{v}_1)_A + (m\mathbf{v}_1)_B = (m\mathbf{v}_2)_A + (m\mathbf{v}_2)_B \qquad (11\text{-}2)$$

where

   m represents the mass of each of the bodies, and

   $\mathbf{v}_1$ and $\mathbf{v}_2$ represent the initial and final velocities respectively of the bodies.

   Equation 11-2 is an expression of the principle of conservation of linear momentum. Once again it must be emphasized that this is a vector equation.

   Since the net external force in the direction of motion in a system such as that shown in Fig. 11-4 is zero, then it follows that the center of gravity of the system will not accelerate even though the bodies making up the system may accelerate toward the center of gravity. In other words, the velocity of the center of gravity of the system must remain constant or zero.

EXAMPLE 11-3

   Two boxcars are moving in the same direction along a track in a railway marshalling yard. If car A weighs 40,000 lb and has a speed of 3 mi/hr and car B weighs 25,000 lb and has a speed of 8 mi/hr, determine the speed of the two cars after they couple together.

This problem may be solved by simple substitution of values in Eq. (11-2), Since all the vectors are in the same direction, we need not be concerned with any more than the sign of each term.

Since the conversion of pounds to slugs is common to each term, it has been eliminated, as has the conversion from miles per hour to feet per second. Since the cars couple together, v_2 must be the same for both cars.

$$(mv_1)_A + (mv_1)_B = (mv_2)_A + (mv_2)_B$$
$$40{,}000 \times 3 + 25{,}000 \times 8$$
$$= 40{,}000v_2 + 25{,}000v_2$$
$$120 + 200 = 40v_2 + 25v_2$$
$$65v_2 = 320$$
$$v_2 = \frac{320}{65} = 4.93 \text{ mi/hr}$$

PROBLEMS

11-1. A horizontal force of 80 lb is applied to a 500-lb block resting on a smooth level surface. If the force is applied for 3 sec and the block starts from rest, find the final speed of the block.

11-2. For an airplane to take off from short runways, small rockets are sometimes attached to the craft to increase the takeoff speed. If an aircraft weighs 18,000 lb and has a speed of 120 mi/hr when the rockets are fired for 8 sec, determine the speed of the aircraft in mi/hr at the end of the rocket firing. The total thrust of the rockets is 5000 lb.

11-3. A 30,000 ton steamship moving at 0.1 mi/hr is about to dock. If it strikes the dock and comes to rest in 10 sec, what is the average force on the dock during this time?

11-4. A 20-ton boxcar at rest on a level track is struck by a 30-ton car traveling at 6 mi/hr. If the speed after impact is 3.6 mi/hr and the impact lasts 0.5 sec, what is the average force on the coupling?

11-5. A gun weighing 160,000 lb fires a 900-lb projectile whose muzzle velocity is 1400 ft/sec. Neglecting the reaction due to

the expanding gases, determine the maximum speed of recoil. What must the elevation be of the gun barrel to achieve this maximum speed?

11-6. A 2500-lb car traveling at 15 mi/hr hits a 3500-lb car stopped at an intersection. If the two cars lock together, and assuming that the brakes are released on both cars, determine the speed of the two cars after impact.

11-7. A 210-lb football player tackles a stationary 400-lb tackling dummy at 15 ft/sec. Determine the speed of the player and dummy immediately after impact.

11-8. A 900-lb ram on a pile driver strikes a 1500-lb pile with a speed of 12 ft/sec. If there is no rebound of the ram from the pile, determine the speed of the ram and pile immediately after impact.

11-9. A bullet weighing 0.75 oz is fired horizontally with a speed of 1800 ft/sec into a 10-lb block, into which it is imbedded. Knowing that the coefficient of friction between the block and floor is 0.30, determine the distance through which the block will slide.

11-3. ANGULAR IMPULSE AND MOMENTUM

The relationship between angular impulse and momentum can be developed by a consideration of Newton's second law as it applies to rotational motion, as follows:

$$T = I\alpha$$

But $\alpha = \Delta\omega/\Delta t$, which can be substituted, so that we have

$$T = I\frac{\Delta\omega}{\Delta t}$$

If we multiply both sides of the equation by Δt we get

$$T\,\Delta t = I\frac{\Delta\omega}{\Delta t}\,\Delta t = I\,\Delta\omega$$

or

$$T\,\Delta t = I(\omega_2 - \omega_1) \tag{11-3}$$

where

     T is the torque applied to the body,
     Δt is the time interval during which the torque is applied,
     I is the mass moment of inertia with respect to the axis
of rotation through the center of gravity of the body,
and ω_1 and ω_2 are the initial and final angular velocities
respectively.
     The units customarily used are ft-lb for torque, sec for
time, slug-ft^2 or lb-ft-sec^2 for moment of inertia, and rad/sec
for angular velocity.
     The angular impulse is $T\,\Delta t$, and the angular momentum
is $I\omega$. Both have units of ft-lb-sec. Although both are vectors,
we generally need not be concerned with anything more than
the signs in plane problems, since all the vectors will usually
be parallel to the z axis for the problems which we will work
with.

EXAMPLE 11-4

     A 1000-lb flywheel is rotating at 650 rpm. It is to be
braked to a stop in 12 sec. Determine the braking torque
required if the radius of gyration of the flywheel is 2.3 ft.

The torque can be found by substituting
appropriate values in Eq. (11-3). It would
probably simplify the numbers if ω in
rad/sec and I are found first.

$$\omega = \frac{650 \times 2\pi}{60}$$
$$= 68.1 \text{ rad/sec}$$

$$I = \frac{1000 \times 2.3^2}{32.2}$$

$$= 164.5 \text{ lb-ft-sec}^2$$

$$T \Delta t = I(\omega_2 - \omega_1)$$

$$T \times 12 = 164.5(0 - 68.1)$$

$$T = -\frac{164.5 \times 68.1}{12}$$

$$= -934 \text{ ft-lb}$$

11-4. CONSERVATION OF MOMENTUM FOR ANGULAR MOTION

The principle of conservation of momentum applies to angular momentum as well as linear momentum. It may be proven by similar arguments to those used in developing the linear relationship in Section 11-2 that the principle of conservation of angular momentum may be expressed as

$$(I\omega_1)_A + (I\omega_1)_B = (I\omega_2)_A + (I\omega_2)_B \qquad (11\text{-}4)$$

where

I is the mass moment of inertia of each of the respective bodies about the axis of rotation, and

ω_1 and ω_2 are the initial and final angular velocities respectively of the bodies.

Although the terms in Eq. (11-4) are vectors, they will all be parallel in our plane problems, so we need to be concerned only with the signs of the terms in the equation.

EXAMPLE 11-5

Two gears in a drive train are brought together to activate the take-off shaft. If the drive gear A is rotating at 450 rpm before meshing with the driven gear B which is initially stationary, find the angular velocity in rpm of the two gears after meshing. The drive gear weighs 300 lb and has a radius of gyration of 1.6 ft and a diameter of 4 ft. The driven gear weighs 120 lb, has a radius of gyration of 1.2 ft and a diameter of 3 ft. The moments of inertia of the shafts are small enough to be neglected.

The final angular velocity may be found by direct substitution of values in Eq. (11-4).

Since the two gears are in contact, we must relate their final angular velocities so that there will be only one unknown term in Eq. (11-4).

A point on the circumference of each gear will have the same final speed. Thus

$$v_A = r_A \omega_A = v_B = r_B \omega_B$$

$$\omega_A = \frac{r_B \omega_B}{r_A}$$

$$= \frac{1.5}{2.0} \omega_B = 0.75 \omega_B$$

To simplify the terms in the equation, it would also help if we calculated the moments of inertia separately.

$$I_A = \frac{300}{32.2} \times 1.6^2 = 23.8 \text{ lb-ft-sec}^2$$

$$I_B = \frac{120}{32.2} \times 1.2^2 = 5.36 \text{ lb-ft-sec}^2$$

Since an ω term is common to each term of the equation, it is not necessary to convert ω to rad/sec.

$$(I\omega_1)_A + (I\omega_1)_B = (I\omega_2)_A + (I\omega_2)_B$$

$$23.8 \times 450 + 0 = 23.8(0.75\omega_B) + 5.36\omega_B$$

$$10,700 = 17.84\omega_B + 5.36\omega_B$$

$$\omega_B = \frac{10700}{23.20} - 461 \text{ rpm}$$

$$\omega_A = 0.75\omega_B = 0.75 \times 461 - 346 \text{ rpm}$$

11-5. COMBINED TRANSLATION AND ROTATION

If a body has motion consisting of a combination of translation and rotation, such as a wheel rolling along a surface, the impulse-momentum principles can still be used to obtain information about the motion of the body. The equation for linear motion must be satisfied for forces through the center of gravity, and the equation of rotation must also be satisfied for torques about an axis through the center of gravity. In other words, both Eq. (11-1) and Eq. (11-3) must be satisfied. This concept is illustrated in the following example.

EXAMPLE 11-6

The 300-lb cylinder shown in Fig. 11-5(a) has a force of 75 lb applied as shown. Determine the speed of the cylinder after 5 sec, if it starts from rest. The coefficient of friction is 0.10.

Step one is to draw a free-body diagram of the cylinder so that all the forces which cause motion can be determined. Since we

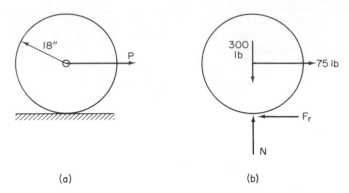

(a)   (b)

Figure 11-5

assume that the cylinder rolls without slipping, the friction force F_r must exist to cause rotation about the center. However, since the cylinder is not necessarily on the verge of slipping, we do not know the size of the friction force.

If we examine our two impulse-momentum equations, we find that we will have three unknowns. However, v_2 and ω_2 can be related, so that we will have three simultaneous equations in three unknowns.

Moment of inertia of cylinder

$$I = \tfrac{1}{2}mr^2$$

$$= \frac{1}{2} \times \frac{300}{32.2} \times 1.5^2$$

$$= 10.5 \text{ lb-ft-sec}^2$$

$$\sum (F \, \Delta t) = mv_2 - mv_1$$

$$(75 - F_r)5 = \frac{300}{32.2} v_2 - 0$$

$$375 - 5F_r = 9.32v_2 \qquad\qquad \text{(a)}$$

$$T \, \Delta t = I\omega_2 - I\omega_1$$

$$1.5F_r \times 5 = 10.5\omega_2 - 0$$

$$7.5F_r = 10.5\omega_2 \qquad\qquad \text{(b)}$$

But

$$\omega_2 = \frac{v_2}{r} = \frac{v_2}{1.5} \qquad\qquad \text{(c)}$$

Substituting in Eq. (b)

$$7.5F_r = 10.5 \times \frac{v_2}{1.5}$$

$$F_r = \frac{10.5v_2}{7.5 \times 1.5}$$

$$= 0.934v_2$$

Substituting in Eq. (a)

$$375 - 5(.934v_2) = 9.32v_2$$

$$375 - 4.67v_2 = 9.32v_2$$

$$v_2 = \frac{375}{13.99} = 26.8 \text{ ft/sec}$$

The size of the friction force must be calculated to determine that the actual friction is less than the maximum possible friction.

From Eq. (b)

$$F_r = \frac{10.5 \times 26.8}{7.5 \times 1.5} = 25 \text{ lb}$$

$$F_{r(\text{max})} - \mu N$$
$$= 0.10 \times 300 = 30 \text{ lb}$$

Thus cylinder rolls without slipping.

PROBLEMS

11-10. A torque of 50 in-lb is applied for 8 sec to a pulley which has a radius of gyration of 0.8 ft and a weight of 240 lb. If the pulley starts from rest, determine the final angular velocity of the pulley in rpm.

11-11. The belt tensions are 95 lb and 18 lb respectively on a 600-lb pulley with a 28-in. diameter. If the radius of gyration of the pulley is 10 in., determine the angular velocity of the pulley in rpm after 4 sec if its initial angular velocity was 90 rpm.

11-12. A billiard ball weighing 5 oz and having a radius of 1.125 in. is struck tangentially with a cue so that it spins without translation. If the force exerted is 1.2 lb and it lasts for 0.2 sec, determine the angular velocity given to the ball.

11-13. The input side of a friction clutch weighs 85 lb and has a radius of gyration of 1.0 ft. The output side weighs 170 lb and has a radius of gyration of 0.8 ft. If the input is operating at 1200 rpm, determine the angular velocity in rpm of the system after the clutch is engaged if the output was initially at rest.

11-14. The two bars shown in Fig. P11-14 are attached to a collar which rotates at 500

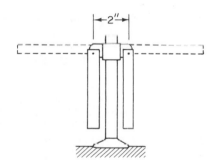

Figure P11-14

rpm. An internal mechanism raises the bars to the horizontal position. Determine the angular velocity of the mechanism in rpm after the bars are raised. The bars each weigh 8 lb and are 14 in. long. The bars initially have a distance of 2 in. between their axes. When they are raised their ends are 1.5 in. apart.

11-15. A torque of 10 ft-lb is applied to a wheel, which causes the wheel to roll without slipping. If the wheel starts from rest, weighs 40 lb, and has a moment of inertia of 4.2 slug-ft², determine its speed after the torque has been applied for 8 sec. The radius is 1.50 ft.

11-16. A pair of disks connected by a hub as shown in Fig. P11-16 weigh 40 lb and have a radius of gyration of 0.9 ft. The 20-lb force causes the disks to roll without slipping. Determine the speed of the center of the disks after 2 sec.

11-17. The solid homogeneous cylinder shown in Fig. P11-17 weighs 200 lb and

has a radius of 1.5 ft. It is released from rest from the position shown. If the rope is wrapped around the cylinder so that it unwinds freely, determine the speed of the cylinder's center 4 sec after release.

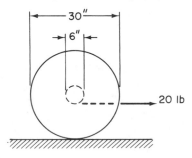

Figure P11-16

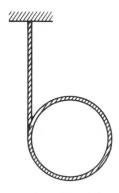

Figure P11-17

11-6. REVIEW OF METHODS OF SOLUTION OF KINETICS PROBLEMS

We have studied the use of three methods to solve problems relating forces and the motion they cause. These are (1) Newton's second law, (2) work and energy, and (3) impulse and momentum. The last two methods were derived from Newton's second law. The advantage in having these different

Table 11–1. Methods of Solution for Kinetics Problems

Method	Variables Used in Solution*
Linear Motion	
Newton's second law	$\mathbf{F}, m, \mathbf{a}$
Work and energy	$\mathbf{F}, m, \mathbf{s}, \mathbf{v}$
Impulse and momentum	$\mathbf{F}, m, \mathbf{v}, t$
Angular Motion	
Newton's second law	$\mathbf{T}, I, \boldsymbol{\alpha}$
Work and energy	$\mathbf{T}, I, \boldsymbol{\theta}, \boldsymbol{\omega}$
Impulse and momentum	$\mathbf{T}, I, \boldsymbol{\omega}, t$

*All terms except m, I, and t are vectors. However, in many of the problem solutions, the scalar magnitudes only will be required.

methods available for problem solution is that different variables are used in the different methods of problem solution. The information used in solving problems employing the different methods is summarized in Table 11-1.

You will find that most of the problems of Chapters 9, 10, and 11 can be solved using any of the methods along with some of the equations of kinematics from Chapter 8. It is highly recommended to review the problems in Chapters 9, 10, and 11 using a method different from the method used in the particular chapter.

11-7. IMPACT

If two bodies are in collision, any of several things may occur after the collision. The two bodies may stick together and continue moving, or they may "bounce" apart or they may stick together without further motion. If they do move away from each other after collision, their relative velocities will depend on the materials of the bodies. Notice that in nearly all discussion up to this point, we have been concerned about rigid bodies. What happens during impact is related to how much kinetic energy is dissipated by deformation of the bodies during impact.

The relative velocities of two bodies A and B before and after impact reflect the loss of kinetic energy. The relative velocities are related as follows:

$$e = \frac{v_{B_2} - v_{A_2}}{v_{A_1} - v_{B_1}} \qquad (11\text{-}5)$$

where

  e is the coefficient of restitution,
  v_{A_2} is the velocity after impact of A,
  v_{B_2} is the velocity after impact of B,
  v_{A_1} is the velocity before impact of A, and
  v_{B_1} is the velocity before impact of B.

The coefficient of restitution is a measure of the elasticity of impact. For a perfectly elastic impact, e has a value of 1, and there is no loss of kinetic energy. If the two bodies stay together after impact, the value for e is zero, and the impact is said to be plastic. For most colliding bodies the value for e will be between 1 and 0.

If the coefficient of restitution between two colliding bodies is known and their initial velocities are also known, there would still be two unknown terms in Eq. (11-5). The second equation required is the conservation of momentum equation, which is Eq. (11-2). Although kinetic energy is usually not conserved in an impact, momentum is conserved, so that Eq. (11-2) and (11-5) may be used together to determine the velocities of the two bodies after impact.

The two most common types of impact problem are illustrated in Fig. 11-6. Direct central impact is shown in Fig.

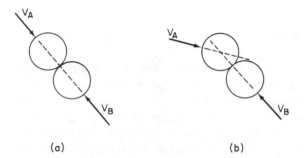

Figure 11-6   (a)   (b)

11-6(a). In this case the centers of gravity move along the same path. Oblique central impact is illustrated in Fig. 11-6(b). The paths of the centers of gravity are not colinear. In the case of oblique central impact, the coefficient of restitution applies to the component of velocity normal to the contact surface only.

EXAMPLE 11-7

A 5-lb ball is rolled along a table top and strikes a second stationary 3-lb ball at 6 ft/sec. If the coefficient of restitution between the two balls is 0.85, determine the speed of each ball after impact.

Values of the velocities are substituted in both the conservation of momentum equation and the coefficient of restitution equation.

$$(mv_1)_A + (mv_1)_B = (mv_2)_A + (mv_2)_B$$

$$\frac{5}{32.2} \times 6 + 0 = \frac{5}{32.2} v_{A_2} + \frac{3}{32.2} v_{B_2}$$

$$30 = 5v_{A_2} + 3v_{B_2} \tag{a}$$

$$e = \frac{v_{B_2} - v_{A_2}}{v_{A_1} - v_{B_1}}$$

$$0.85 = \frac{v_{B_2} - v_{A_2}}{6 - 0}$$

$$5.10 = v_{B_2} - v_{A_2}$$

$$v_{A_2} = v_{B_2} - 5.10 \tag{b}$$

Substituting in Eq. (a)

$$30 = 5(v_{B_2} - 5.10) + 3v_{B_2}$$

$$30 = 5v_{B_2} - 25.5 + 3v_{B_2}$$

$$8v_{B_2} = 55.5$$

$$v_{B_2} = \frac{55.5}{8} = 6.94 \text{ ft/sec}$$

From Eq. (b)

$$v_{A_2} = 6.94 - 5.10 = 1.84 \text{ ft/sec}$$

PROBLEMS

11-18. Two identical bodies move along the same path. If A travels at 8 ft/sec and B at 17 ft/sec, determine the speeds of A and B after impact if the coefficient of restitution is 0.75.

11-19. If the coefficient of restitution of the collars shown in Fig. P11-19 is 0.60, deter-

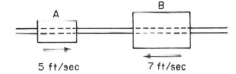

B

A

5 ft/sec   7 ft/sec

Figure P11-19

mine the velocity of each collar after impact. A weighs 6 lb, and B weighs 9 lb.

11-20. Two balls A and B have initial velo-

cities of 4 ft/sec and -3 ft/sec, respectively. Their final velocities after impact are 0.5 ft/sec and 2.0 ft/sec, respectively. What is the coefficient of restitution between the two bodies?

11-21. A ball weighing 1 lb is dropped onto the floor from a height of 4 ft. If the coefficient of restitution between the floor and the ball is 0.85, determine how high the ball will bounce. Determine the height of the second bounce.

11-22. A 12-lb body moves in the positive x direction, and a 7-lb body moves in the positive y direction. If both have identical speeds of 20 ft/sec, determine their velocities after impact if the coefficient of restitution is 0.65.

ANSWERS TO EVEN NUMBERED QUESTIONS

These answers have been obtained using a slide rule, so minor differences are to be expected. Directions given in the answers are positive (counterclockwise) angles measured from the positive x axis.

2-2 (a) $r = 2.33$ in.
  (b) $x = 1.79$.
  (c) $\theta = 50°$

2-4 $C_1 = 11.74$ in.
  $\beta_1 = 38.8°$
  $\gamma_1 = 113.2°$
  $C_2 = 2.40$ in.
  $\beta_2 = 141.2°$
  $\gamma_2 = 10.8°$

2-6 $\alpha = 46.6°$
  $\beta = 58.0°$
  $\gamma = 75.4°$

2-8 $C = 12.0$ ft
  $\alpha = 42.8°$
  $\beta = 22.2°$

2-10 $AB = 366$ ft

2-12 Guy $= 447$ ft
  Distance $= 259$ ft

3-2 $R = 17.33$ units, $0°$

3-4 $R = 23.8$ units, $82.5°$
  $R = 18.26$ units, $172.6°$

3-6 $d_E = 1.64$ mi
  $d_N = 1.15$ mi

3-8 433 lb, $330°$
  250 lb, $60°$

3-10 $C = 221$ lb, $210°$

3-12 $R = 34.7$ units, $39.2°$

3-14 $v = 65.0$ ft/sec, $22.6°$

327

3-16 $R = 95.4$ lb, $349.9°$

3-18 $R = 45.7$ lb, $217.0°$

3-20 $R = 1.23$ k, $234.2°$

3-22 (a) $M_A = 480$ in-lb
$\qquad\qquad M_B = -412$ in-lb

3-24 $M_C = -2.0$ ft-k

3-26 $M_A = 450,000$ ft-lb

3-28 $C = 600$ in-lb

3-30 $C = 1200$ in-lb

3-32 $C = 30$ in-lb

3-34 800 lb down, 1.50 ft right of A

3-36 4.0 k down, 7.75 ft right of A

3-38 42 tons, 20.2 ft behind front wheels

3-40 $R = 339$ lb, $247.6°$, 2.44 ft right of A

3-42 $R = 1.197$ k, $135.4°$, 9.90 ft right of A

3-44 $R = 44,200$ lb, $187.6°$, 161 ft ahead of bow

3-46 $F = 50$ lb to the left
$\qquad\quad C = 150$ in-lb

3-48 $R = 20$ k down
$\qquad\quad C = -168$ in-k

3-50 $R = 31.5$ lb, $34.0°$
$\qquad\quad C = 12.2$ in-lb

4-2 $F = 65.8$ lb, $247.0°$

4-4 $T_{AC} = 2.64$ k
$\qquad T_{BC} = 1.95$ k

4-6 $R_A = 742$ lb, $339°$
$\qquad R_B = 666$ lb, up

4-8 $T_{AB} = 323$ lb, $148.2°$

4-10 $W = 561$ lb
$\qquad\quad \alpha = 75.5°$

4-12 $R_A = 19.22$ lb, $69.5°$
$\qquad\quad R_B = 6.73$ lb, $180.0°$

4-14 $R_A = 5.33$ k, up
$\qquad\quad R_B = 8.67$ k, up

4-16 $R_A = 0.86$ k, down
$\qquad\quad R_B = 10.86$ k, up

4-18 $R_A = 5.49$ k, $160.9°$
$\qquad\quad R_B = 1.20$ k, up

4-20 $R_A = 0.18$ k, down
$\qquad\quad R_B = 3.04$ k, $311.2°$

4-22 $R_A = 0.375$ k, down
$\qquad\quad R_B = 0.375$ k, up

4-24 $R_A = 1050$ lb, $250°$
$\qquad\quad R_B = 1470$ lb, up

4-26 $R_A = 16.67$ lb, down
$\qquad\quad R_B = 16.67$ lb, up

4-28 $R_A = 100$ lb, left
$\qquad\quad R_B = 50$ lb, up

4-30 $d = 57.5$ ft

4-32 $R_A = 1.46$ k, $233.2°$
$\qquad\quad R_B = 1.46$ k, $233.2°$
$\qquad\quad R_C = 3.34$ k, $159.5°$

4-34 $A_x = 1.00$ k, right
$\qquad\quad A_y = 0.75$ k, up
$\qquad\quad R_D = 1.25$ k, $323.2°$

4-36 $A_x = 90.9$ lb, right
$\qquad\quad A_y = 150.9$ lb, up
$\qquad\quad B_x = 109.1$ lb, right
$\qquad\quad B_y = 50.9$ lb, down

4-38 $R = 120$ lb

4-40 $R_A = 6.94$ k, $133.9°$

4-42 $R_A = 3.66$ k, $82.5°$

4-44 $F = 215$ lb, $62.2°$

4-46 $T_{AB} = 519$ lb
$\qquad\quad D_x = 1200$ lb
$\qquad\quad D_y = 693$ lb

4-48 $T_{AB} = 3.85$ k
$\qquad\quad R_C = 4.14$ k, $71.3°$

4-50 $F = 47.1$ k

4-52 $T_{AB} = 3.14$ k
$\qquad\quad R_D = 1.59$ k, $10.1°$

4-54 $F_{BD} = 1.30$ k

4-56 $F_{DB} = -333$ lb
$\qquad\quad A_x = 667$ lb right
$\qquad\quad A_y = 1000$ lb up

4-58 $F_{AB} = -1.485$ k
$\quad\quad F_{BC} = -1.37$ k
$\quad\quad F_{AG} = 2.05$ k

4-60 $F_{AB} = -4.24$ k
$\quad\quad F_{BC} = -0.85$ k
$\quad\quad F_{CE} = -6.00$ k
$\quad\quad F_{EF} = -2.25$ k
$\quad\quad F_{FG} = 3.00$ k
$\quad\quad F_{GA} = 3.00$ k
$\quad\quad F_{BG} = 0$
$\quad\quad F_{BF} = -3.07$ k
$\quad\quad F_{CF} = 6.90$ k

4-62 $F_{DE} = 0$
$\quad\quad F_{DG} = -8.04$ k
$\quad\quad F_{CD} = 3.86$ k
$\quad\quad F_{CG} = -2.96$ k

4-64 $R_A = 6.03$ k, 204.2°
$\quad\quad R_G = 10.8$ k, up
$\quad\quad F_{GH} = 1.22$ k
$\quad\quad F_{CG} = 0$
$\quad\quad F_{GJ} = -9.94$ k

4-66 $F_{RF} = 10.0$ k
$\quad\quad F_{KG} = 28.6$ k
$\quad\quad F_{DH} = 0$

4-68 $F_{BF} = 0$
$\quad\quad F_{BE} = -10.8$ k
$\quad\quad F_{CD} = -10.4$ k

4-70 $F_{AB} = 8.66$ k
$\quad\quad F_{CD} = -10.0$ k

4-72 $F_{AB} = -378$ lb

4-74 $F_{CF} = -1.55$ k
$\quad\quad F_{HG} = 6.04$ k

4-76 $P = 42.4$ k
$\quad\quad F_{BH} = 0$

4-78 $P = 7.5$ k

5-2 $F_x = 285$ lb
$\quad\quad F_y = 342$ lb
$\quad\quad F_z = 228$ lb

5-4 cos $\theta_x = -0.667$
$\quad\quad$ cos $\theta_y = 0.667$
$\quad\quad$ cos $\theta_z = 0.333$
$\quad\quad F_x = -600$ lb

$\quad\quad F_y = 600$ lb
$\quad\quad F_z = 300$ lb

5-6 $W = 35.1$ k

5-8 $C_x = -768$ ft-lb
$\quad\quad C_y = -512$ ft-lb
$\quad\quad C_z = -384$ ft-lb
$\quad\quad$ cos $\theta_x = -0.768$
$\quad\quad$ cos $\theta_y = -0.512$
$\quad\quad$ cos $\theta_z = -0.384$

5-10 $R = 99.2$ lb
$\quad\quad$ cos $\theta_x = -0.204$
$\quad\quad$ cos $\theta_y = 0.980$
$\quad\quad$ cos $\theta_z = 0.0262$

5-12 $R = 386$ lb

5-14 $R = 688$ lb
$\quad\quad$ cos $\theta_x = 0.673$
$\quad\quad$ cos $\theta_y = 0.605$
$\quad\quad$ cos $\theta_z = -0.425$

5-16 $R = 20.0$ k, down
$\quad\quad \bar{x} = 4.95$ ft
$\quad\quad \bar{y} = 6.75$ ft

5-18 $R = 5.00$ k, up
$\quad\quad \bar{x} = 2.40$ ft
$\quad\quad \bar{y} = 9.60$ ft

5-20 $\bar{x} = 1.50$ ft
$\quad\quad \bar{z} = 1.00$ ft

5-22 $T = 208$ lb

5-24 $F_{AD} = 561$ lb
$\quad\quad F_{BD} = 561$ lb
$\quad\quad F_{CD} = -1178$ lb

5-26 $T = 5.56$ k

5-28 $M_{aa} = 43.3$ ft-lb

5-30 $P = 20$ lb
$\quad\quad A_y = -30$ lb
$\quad\quad A_z = 50$ lb
$\quad\quad B_y = 10$ lb
$\quad\quad B_z = 50$ lb

5-32 $R = 64.0$ lb
$\quad\quad \theta_x = 128.6°$
$\quad\quad \theta_y = 141.4°$
$\quad\quad \theta_z = 90°$

$C = 889$ in-lb
$\theta_x = 66.8°$
$\theta_y = 148.8°$
$\theta_z = 70.2°$

6-2 $F = 30$ lb

6-4 $\mu = 0.70$

6-6 $W = 26.9$ lb

6-8 $N = 379$ lb

6-10 $P = 26.2$ lb

6-12 $W_A = 2.96$ k

6-14 $P = 746$ lb

6-16 $P = 75.0$ lb

6-18 (a) $R_W = 160$ lb
      (b) $\mu = 0.575$

6-20 $P = 100$ lb

6-22 Lever: $P = 333$ lb
      Wedge: $P = 261$ lb

6-24 $P = 77.0$ lb

6-26 $P = 678$ lb

6-28 $T = 72.0$ in-lb

6-30 $T = -5.16$ in-lb

6-32 $F = 1.93$ lb

6-34 (a) $T_2 = 89.0$ lb
      (b) $T = 244$ in-lb

6-36 $T_2 = 161$ lb

6-38 $W = 577$ lb

7-2 $\bar{x} = 1.25$ in.
     $\bar{y} = 4.00$ in.

7-4 $\bar{x} = -1.71$ in.
     $\bar{y} = -0.292$ in.

7-6 $\bar{x} = 3.37$ in.
     $\bar{y} = 6.95$ in.
     $\bar{z} = 1.68$ in.

7-8 $\bar{x} = 5.26$ in.
     $\bar{y} = 0.574$ in.
     $\bar{z} = -1.22$ in.

7-10 $\bar{x} = 4.90$ in.

7-12 $\bar{z} = 3.44$ in.

7-14 $\bar{x} = 0$
      $\bar{y} = 0$
      $\bar{z} = 3.15$ ft

7-16 $\bar{y} = 3.07$ in.

7-18 $\bar{x} = 2.50$ in.
      $\bar{y} = 4.00$ in.

7-20 $\bar{x} = 3.71$ in.

7-22 $\bar{y} = 2.74$ in.

7-24 $\bar{y} = 5.75$ in.

7-26 $\bar{x} = 0$
      $\bar{y} = 6.98$ in.

7-28 $\bar{x} = 0$
      $\bar{y} = 0.516$ in.

7-30 $\bar{z} = 5.03$ in.

7-32 $\bar{x} = 1.00$ ft
      $\bar{y} = -1.90$ ft
      $\bar{z} = 4.00$ ft

7-34 $W = 8000$ lb, 10 ft from end

7-36 $W = 860$ lb, 5.61 ft from left end

7-38 $R_A = 256$ lb, up
      $R_B = 64.0$ lb, up

7-40 $F = 283,000$ lb
      $M = 158,000,000$ ft-lb

7-42 $A = 223$ in.^{2}

7-44 $V = 1143$ in.^{3}
      $A = 495$ in.^{2}

7-46 $V = 0.348$ in.^{3}

7-48 $I_x = 130$ in.^{4}

7-50 $I_y = 2210$ in.^{4}

7-52 $I_x = 41.7$ in.^{4}

7-54 $k_x = 2.74$ in.

7-56 $I_x = 7770$ in.^{4}

7-58 $I_x = 1272$ in.^{4}

7-60 $I_y = 16.0$ in.^{4}

7-62 $I_x = 530$ in.^{4}

7-64 $A = 44.4$ in.^{2}

7-66 $I_y = 585$ in.^{4}

7-68 $I_x = 164.5$ in.^{4}
$\bar{I}_x = 32.5$ in.^{4}

7-70 $\bar{I}_x = 741$ in.^{4}

7-72 $I_x = 242$ in.^{4}

7-74 $\bar{x} = 1.00$ in.
$\bar{y} = -0.333$ in.
$\bar{I}_x - 181.3$ in.^{4}

7-76 $k_{AA} = 5.16$ in.

7-78 $\bar{J} - 300$ in.^{4}
$J = 400$ in.^{4}

7-80 $J = 62$ in.^{4}

7-82 $J = (3/2)\pi r^4$

7-84 $I = 31.2$ slug-ft^2

7-86 $k_y = 3.24$ ft

7-88 $I_z = 1.11$ slug-ft^2

7-90 $I = 49.8$ slug-ft^2

7-92 $I_z = 1.91$ slug-ft^2

8-2 $d_i = 2.0$ ft, $90°$
$d_f = 2.0$ ft, $270°$
distance $= 6.28$ ft

8-4 $v = 15$ ft/sec, S $20°$ E

8-6 $v = 7.85$ ft/sec, $91.5°$
$v = 7.07$ ft/sec, $135.0°$

8-8 $a = 94.4$ ft/sec^2, $148.0°$
$a = 30,000$ mi/hr^2, $148.0°$

8-10 $a = 67.3$ ft/sec^2, $76.3°$

8-12 $a = 0.454$ ft/sec^2, $233.5°$

8-14 $t = 2.49$ sec
$t = 3.52$ sec

8-16 Yes $(s = 56.0$ ft)

8-18 $t = 0.558$ sec
$s = 836$ ft

8-20 $v = 56.8$ ft/sec

8-22 $v = 30$ ft/sec (constant)

8-24 $a = 6$ mi/hr^2 (constant)

8-26 For $t = 0$ to 15 sec, $v = 0.533$ ft/sec
For $t > 15$ sec, $v = -1.40$ ft/sec

8-28 For $t = 10$ sec, $v = 0.80$ in/sec, $a = 0.092$ in/sec^2
For $t = 20$ sec, $v = 2.2$ in/sec, $a = 0.260$ in/sec^2
For $t = 25$ to 40 sec, $v = 0.67$ in/sec, $a = 0$ in/sec^2

8-30 For $t = 1.0$ sec, $v = 1.92$ in/sec, $a = 1.80$ in/sec^2
For $t = 2.0$ sec, $v = -1.60$ in/sec, $a = -1.20$ in/sec^2
For $t = 3.0$ sec, $v = 2.16$ in/sec, a is undefined

8-32 For $t = 4.8$ hr, $s = -96$ mi, $a = 8.33$ mi/hr^2
For $t = 12$ hr, $s = 120$ mi, $a = 8.33$ mi/hr^2

8-34 For $t = 8$ sec, $v = 80$ ft/sec, $s = 320$ ft
For $t = 18.33$ sec, $v = 0$, $s = 933$ ft

8-36 For $t - 25$ sec, $v = 900$ ft/sec, $s = 18,300$ ft
For $t = 45$ sec, $v = 900$ ft/sec, $s = 36,300$ ft
For $t = 75$ sec, $v = 1125$ ft/sec, $s = 65,600$ ft

8-38 For $t = 0.2$ sec, $v = 0.5$ in/sec, $s = 2.047$ in.
For $t = 0.35$ sec, $v = 0.95$ in/sec, $s = 2.156$ in.

8-40 $v = 9.16$ ft/sec, $240°$

8-42 $v = 14.14$ ft/sec, $240°$

8-44 (a) $a_n = 4.30$ ft/sec^2
(b) $a_n = 2.10$ ft/sec^2

8-46 $\omega = 740$ rpm

8-48 $v = 34.9$ ft/sec
$v = 104.6$ ft/sec
$v = 174.5$ ft/sec

8-50 $v = 26.2$ ft/sec

8-52 $v = 26.2$ ft/sec
$a = 302$ ft/sec^2

8-54 $v = 5.0$ ft/sec, $60°$
$a = 104.2$ ft/sec^2, $133.3°$

8-56 $\alpha = 200$ rad/sec^2

8-58 $\alpha = 60.1$ rad/sec^2

8-60 $t = 5.26$ sec

8-62 $\theta = 50°$

8-64 $\omega = 112.5$ rpm

8-66 $\alpha = -2.40$ rad/sec^2

8-68 36 in., 27 in., 13.5 in.

8-70 $\omega = 16,000$ rpm

8-72 $\alpha = -51.7$ rad/sec^2

8-74 $t = 3.28$ sec

8-76 $v = 1.5$ ft/sec

8-78 $s_{c/t} = 0.463$ mi, 157.6°

8-80 Course N 76.8° E
$\quad\quad v = 102.1$ mi/hr

8-82 $v = 15.46$ in/sec, 195°

8-84 $v = 8.0$ ft/sec, right

8-86 $v = 80.8$ ft/sec, 333.7°

8-88 $\omega = 22.2$ rad/sec

8-90 $a = 37.8$ ft/sec^2, 238.0°

8-92 $a_B = 4$ ft/sec^2
$\quad\quad a_C = 11$ ft/sec^2
$\quad\quad a_D = 3$ ft/sec^2
$\quad\quad \alpha_{AB} = 0$
$\quad\quad \omega_{AB} = 1.155$ rad/sec
$\quad\quad v_B = 3.47$ ft/sec

9-2 $a = 3.22$ ft/sec^2

9-4 $a = 29.0$ ft sec^2

9-6 $R = 281$ lb

9-8 $a = 9.71$ ft/sec^2

9-10 $a = 13.84$ ft/sec^2

9-12 $a = 3.80$ ft/sec^2, down

9-14 $r = 96,100$ ft

9-16 $\omega = 119$ rpm

9-18 $\omega = 2.93$ rad/sec

9-20 $\omega = 18.3$ rpm

9-22 $a = 11.29$ ft/sec^2

9-24 $\theta = \tan^{-1}(v^2/gr)$

9-26 $a = 16.58$ ft/sec^2

9-28 $\alpha = 107.5$ rad/sec^2

9-30 $W = 14,200$ lb

9-32 $t = 1.89$ sec

9-34 $T_A = 444$ lb
$\quad\quad T_B = 522$ lb

9-36 $\alpha = 2.89$ rad/sec^2

9-38 $v = 16.03$ ft/sec

10-2 $U = 40$ ft-lb

10-4 $U = 62.5$ ft-lb

10-6 $U = 1800$ ft-lb
$\quad\quad U_{net} = 760$ ft-lb

10-8 $U = 600$ ft-lb

10-10 $U = 4500$ ft-lb
$\quad\quad U_{net} = 1147$ ft-lb

10-12 $U = 35.1$ ft-lb

10-14 $U = 200,000$ ft-lb

10-16 $U = 410$ ft-k

10-18 (a) $U = 84$ in-lb
$\quad\quad$ (b) $U = 60$ in-lb

10-20 $U_{net} = 510$ in-lb

10-22 $v = 62.2$ ft/sec

10-24 $v = 14.92$ ft/sec

10-26 $F = 378$ lb

10-28 $v = 19.24$ ft/sec

10-30 $v = 5.16$ ft/sec

10-32 $U = 660,000$ in-lb

10-34 $U = 2830$ ft-lb

10-36 $T = 867$ ft-lb

10-38 $T = 18.02$ ft-lb

10-40 $v = 43.4$ ft/sec
$\quad\quad \omega = 21.7$ rad/sec

10-42 $v = 17.1$ ft/sec

10-44 $P = 40.4$ hp

10-46 $P = 1520$ watts
$\quad\quad P = 2.02$ hp

10-48 Efficiency $= 63.6\%$

10-50 $P = 160$ hp

11-2 $v = 169$ mi/hr

11-4 $F = 13,120$ lb

11-6 $v = 6.25$ mi/hr

11-8 $v = 4.50$ ft/sec

11-10 $\omega = 66.8$ rpm

11-12 $\omega = 660$ rad/sec

11-14 $\omega = 6.55$ rpm

11-16 $v = 17.0$ ft/sec

11-18 $v_A = 15.88$ ft/sec
$v_B = 9.13$ ft/sec

11-20 $e = 0.214$

11-22 $v_A = 14.45$ ft/sec, $57.1°$
$v_B = 20.8$ ft/sec, $357.8°$

INDEX